Spatial Analysis, GIS, and Remote Sensing Applications in the Health Sciences

Editors

Donald P. Albert
Wilbert M. Gesler
Barbara Levergood

CRC Press
Taylor & Francis Group
Boca Raton London New York

CRC Press is an imprint of the
Taylor & Francis Group, an **informa** business

First published 2000 by Ann Arbor Press

Published 2020 by CRC Press
Taylor & Francis Group
6000 Broken Sound Parkway NW, Suite 300
Boca Raton, FL 33487-2742

First issued in paperback 2020

ISBN 13: 978-0-367-57894-7 (pbk)
ISBN 13: 978-1-57504-101-8 (hbk)

**Visit the Taylor & Francis Web site at
http://www.taylorandfrancis.com**

**and the CRC Press Web site at
http://www.crcpress.com**

Library of Congress Cataloging-in-Publication Data

    Spatial analysis, GIS and remote sensing : applications in the health sciences / edited by Donald P. Albert, Wilbert M. Gesler, Barbara Levergood.
        p.   cm.
    Includes bibliographical references and index.
    ISBN 1-57504-101-4
    1. Medical geography. 2. Medical geography—Research—Methodology. I. Albert, Donald Patrick. II. Gesler, Wilbert M., 1941– . III. Levergood, Barbara.
    RA792 .S677 2000
    614.4'2—dc21   99-089917

To

Julie, Elizabeth, and Kenny

Acknowledgments

The editors would like to express their appreciation to Lesa Strikland with Medical Media, VA Medical Center (Durham, North Carolina), Department of Veterans Affairs for her assistance in scanning figures and maps.

About the Authors

Donald P. Albert, Ph.D., is an Assistant Professor in the Department of Geography and Geology at Sam Houston State University in Huntsville, Texas. His interests include applications of geographic information systems within the context of medical geography, health services research, and law enforcement.

Kelly A. Crews-Meyer, Ph.D., is a recent graduate of the University of North Carolina at Chapel Hill and an Assistant Professor of Geography at the University of Texas at Austin. Her current work in population-environment interactions draws upon previous research experience in state government, consulting, and university settings in landuse/landcover change, geographic accessibility, and decision-making as applied to environmental policy and valuation. Her educational background includes a B.S. in Marine Science and a M.A. in Government and International Studies, both from the University of South Carolina, as well as a Masters Certificate in Public Policy Analysis from the University of North Carolina at Chapel Hill.

Charles M. Croner, Ph.D., is a geographer and survey statistician with the Office of Research and Methodology, National Center for Health Statistics, Centers for Disease Prevention and Control (CDC). His research interests are in the use of GIS for disease prevention and health promotion planning, small area analysis, and human visualization and cognition. He is Editor of the widely circulated bimonthly report "Public Health GIS News and Information" (free by request at cmc2@cdc.gov).

Rita Fellers, Ph.D. Student, Department of Geography, University of North Carolina at Chapel Hill. Rita Fellers is a medical geographer with a particular interest in potentially environmentally related diseases such as cancer, and in statistical techniques that improve the quality of information that ecologic studies can produce.

Wilbert Gesler, Ph.D., Dr. Wilbert Gesler is a Full Professor of Geography at the University of North Carolina in Chapel Hill. His major research interests are in the Geography of Health, including studies of accessibility to health care in rural areas, socio-spatial knowledge networks involved in prevention of chronic diseases, and places which have achieved a reputation for healing.

Ron D. Horner, Ph.D., Director, Epidemiologic Research and Information Center at Durham, North Carolina. His research interests are in racial/ethnic and rural/urban variations in the patterns of care for cerebrovascular disease.

Barbara Levergood, Ph.D., Electronic Document Librarian, University of North Carolina at Chapel Hill. Her interests include providing public access to Federal information products in electronic media, statistical data, and geographic information systems.

Joseph Messina, Ph.D. Student, Department of Geography, University of North Carolina at Chapel Hill. He served in the U.S. Army using battlefield GIS to support indirect fire control missions. He worked as a GIS Applications Specialist for the SPOT Image Corporation. While with SPOT, he assisted in the development of the GeoTIFF format, developed new products and remote sensing algorithms, and served as contributing technical editor for SPOTLight magazine. He holds degrees in Biology and Geography from George Mason University.

Peggy Wittie, a medical geographer and GIS specialist, is a doctoral candidate at the University of North Carolina at Chapel Hill and GIS Coordinator for North Carolina Superfund. Her research integrates GIS techniques to study health care access, environmental health and environmental justice issues.

Preface

This book is an expression of the myriad ways in which the range of geo-spatial methods and technologies can be applied to the analysis of issues related to human and environmental health. Since the study and management of the many diverse issues related to human health is one of the most important aspects of human endeavor it is not surprising that it has been a fruitful area for application of geo-spatial analysis tools. Contributions to this book run the gamut of these diverse applications areas from more classical medical geography to the study of infectious disease to environmental health. The tools used in these studies are also diverse—ranging from GIS as a core and unifying technology to geo-spatial statistics and the computer processing of remotely-sensed imagery.

This book should prove useful for practitioners and researchers in the health care and allied fields as well as geographers, epidemiologists, demographers, and other academic researchers. Today one sees a continual increase in the power and ease of use of GIS, better integration and easier availability of related technologies, such as remote sensing and global positioning systems and rapidly falling costs of platforms, peripherals, and programs. Thus, one now sees an increasingly large cadre of users of geo-spatial technology in all fields, including health related ones. The methods and examples provided in this work are a starting point for this growing group of users who will find the power of spatial analysis tools and the increasing availability of data sources to enable them to obtain answers and to arrive at solutions to a host of critical health care related issues. The tools and knowledge are readily available and the skills can be developed by any dedicated user; therefore, what direction users of GIS in health related fields choose to take this and related technologies is now primarily limited by their imaginations.

Dr. Mark R. Leipnik, Ph.D.
Director GIS Laboratory,
Texas Research Institute for
 Environmental Studies,
Assistant Professor,
Department of Geography and Geology,
Sam Houston State University
Huntsville, Texas

Contents

Spatial Analysis, GIS, and Remote Sensing Applications in the Health Sciences

Chapter One

Introduction

Medical geography is a very active subdiscipline of geography which has traditionally focused on the spatial aspects of disease ecology and health care delivery. Until fairly recently, as was the case with most other geographic fields of study, medical geographers collected and analyzed their data using methods such as making on-the-ground observations (e.g., of malarial mosquito habitats) and drawing maps (e.g., of hospital catchment areas) by hand. With the advent of geographic information systems (GIS) and remote sensing (RS) technologies, computers which could handle large amounts of data, and sophisticated spatial analytic software programs, medical geography has been transformed. It is now possible, for example, to make many measurements from far above the earth's surface and produce dozens of maps of disease and health phenomena in a relatively short time. This explosion of new capabilities, however, needs to be systematically organized and discussed so that researchers in medical geography can get to know what resources are now available for their use. In this book we set out to accomplish that task of organization and description.

This volume represents an effort to collect, conceptualize, and synthesize research on geomedical applications of spatial analysis, geographic information systems, and remote sensing. Our purpose is to present a resource guide that will facilitate and stimulate appropriate use of geographic techniques and geographic software (geographic information systems and remote sensing) in health-related issues. Our target audience includes health practitioners, academicians (students and instructors), administrators, departments, offices, institutes, centers, and other health-related organizations that wish to explore the interface between health/disease and spatial analysis, geographic information systems, and remote sensing.

This chapter first sets out the scope of this volume using definitions of geotechniques and health science disciplines. The definitions provide parameters used to determine whether to include or exclude articles for our review. The editors and authors apologize up front for omissions; however, due to space (as well as human) limitations some interesting research might fail to appear in this volume. Second, this chapter describes the annual output of the published research using a basic diffusion model. The model describes stages in the rate of growth of phenomena (i.e., output of research publications) over time. The progression is one that follows from innovation, early majority, late majority, and laggard stages of the diffusion process. Finally, this chapter outlines the organization of the volume; included also is a brief abstract of each chapter.

DEFINITIONS

This volume limits its review of research to studies that have interfaced geotechniques (spatial analysis, geographical information systems, and remote sensing) with health and disease topics. Although two of the editors and several of the contributing authors are medical geographers, studies summarized in this volume emanate not only from medical geography, but also biostatistics, environmental health, epidemiology, health services research, medical entomology, public health, and other related disciplines.

Defining terms is problematic because complementary and contradictory definitions often compete for supremacy or acceptance. Of the three geotechniques, the least definable is GIS. One of the major critiques of GIS is the absence of a universally accepted definition. Fortunately, the eclectic scope of this volume permits the editors to accept the full definitional spectrum of GIS. One might view spatial analysis, GIS, and remote sensing as converging rather than distinct techniques and technologies. For the moment, however, note the following definitions of spatial analysis, GIS, and remote sensing.

Geotechniques

Spatial Analysis: The study of the locations and shapes of geographic features and the relationships between them (Earth Systems Research Institute, 1996).

Geographic Information Systems: ...computer databases that store and manipulate geographic data (Aronoff, 1989).

Remote Sensing: ...imagery is acquired with a sensor other than (or in addition to) a conventional camera through which a scene is recorded, such as by electronic scanning, using radiation outside the normal visual range of the film and camera—microwave, radar, infrared, ultraviolet, as well as multispectral, special techniques are applied to process and interpret remote sensing imagery for the purpose of producing conventional maps, thematic maps, resource surveys, etc., in the fields of agriculture, archaeology, forestry, geology, and others (Campbell, 1987, p. 3).

Interfacing Disciplines

In recent years the use of geotechniques, especially GIS, has been diffusing into the private and public sectors and across disciplines (e.g., city and regional planning, transportation, government, and marketing). This is no less true for disciplines that have health and/or disease as their foci. Some of the disciplines exploring the use of GIS/RS include biostatistics, epidemiology, environmental health, health services research, medical entomology, medical geography, and public health. Definitions of these disciplines are presented below. Again, as with the definition of geographic information systems, there exist complementary, contradictory, and competing statements that define these disciplines.

However, for the purposes of providing a broad-based review of geomedical/ geotechnical applications, the definitions set out below were deemed to be adequate. Each of these disciplines offers a distinct set of knowledge, methods, and approaches; note, however, that there is a substantial overlap among these sciences.

Biostatistics: The science of statistics applied to biological or medical data (*Illustrated Stedman's Medical Dictionary*, 1982, p. 172).

Environmental Health: ... includes both the direct pathological effects of chemical, radiation and biological agents, and the effects (often indirect) on health and well-being of the broad physical, psychological, social and aesthetic environment, which includes housing, urban development, land use and transport (World Health Organization, 1990).

Epidemiology: The study of the prevalence and spread of disease in a community (*Illustrated Stedman's Medical Dictionary*, 1982, p. 474).

Health Services Research: The central feature of health services research is the study of the relationships among structures, processes, and outcomes in the provision of health services (White et al., 1992, p. xix).

Medical Geography: The application of geographical concepts and techniques to health-related problems (Hunter, 1974, p. 3).

Medical Entomology: Zoology which deals with insects that cause disease or serve as vectors of microorganisms that cause disease in man (*Dorland's Illustrated Medical Dictionary*, 1985, p. 448).

Public Health: The art and science of community health concerned with statistics, epidemiology, hygiene, and the prevention and eradication of epidemic diseases (*Illustrated Stedman's Medical Dictionary*, 1982, p. 622).

Together, the interface between geotechniques (spatial analysis, GIS, and remote sensing) and some specific disciplines (biostatistics, epidemiology, environmental health, health services research, medical entomology, medical geography, and public health) sets our parameter limits. The intersection among the three geotechniques and seven disciplines produces a scope for this volume that is wide and inclusive rather than narrow and exclusive.

DIFFUSION OF GEOGRAPHIC TECHNOLOGIES USED IN THE HEALTH SCIENCES

Spatial analysis came to the fore during the "Quantitative Revolution" of the 1960s and 1970s. The linkages between health/disease with GIS/RS began with just a smattering of interest in the 1980s. For the most part, geomedical applications of GIS/RS are a phenomenon of the 1990s. The standard geographic diffusion model provides a means to track the conception and development of geomedical GIS/RS applications research. This model describes diffusion in terms of the number of adopters of an innovation (i.e., publications) over some time period.

There was just a small number of publications through 1990. From 1991 to 1994 the number of publications hovered around two dozen per year. The number of publications continued to increase each year between 1995–1997. From a diffusion standpoint, research output originated in the late 1980s and 1990 (stage 1) and moved into early expansion (stage 2) from 1991–1997. Our suspicion is that research output will remain in the early expansion stage for several more years before entering the late expanding stage (stage 3) of the diffusion process. Further, it will be a decade or more before saturation sets in (stage 4) and the diffusion process is completed and geographic information systems and remote sensing become standard technologies in the investigation of issues of health and disease.

AN OVERVIEW OF THE TEXT

This book contains nine chapters, a master geographic information systems/ remote sensing bibliography, a glossary, and subject and geographical indices.

The next seven chapters (2–8) provide reviews of geomedical applications of spatial analysis (Chapter 2), geographic information systems (Chapters 3–6), and remote sensing (Chapter 7 and 8). Each of these core chapters uses a concept as an organizational theme from which to "hang" existing research. Chapter 2 uses points, lines, areas, and surfaces, or dimensions 0, 1, 2, and 3 respectively, to organize research incorporating spatial analysis and medical geography. Chapters 3 through 6 present specific applications of geographic information systems in medical geography (Chapter 3), health services research (Chapter 4), environmental and public health (Chapter 5) and infectious diseases (Chapter 6). Chapter 3 places articles of interest to medical geographers into one of four basic literature groups (potential, caution, preliminary, and application). Chapter 4 assesses the contribution of geographic information systems to health services research using a four-group classification of operations and functions of geographic information systems software (Aronoff 1989). The focus of Chapter 5 is on infectious diseases and GIS. There are two conceptual themes operating within Chapter 5. First, each of the five infectious diseases discussed (dracunculiasis, babesiosis, Lyme disease, LaCrosse encephalitis, and malaria) is placed within the context of its geographic distribution and current infection trends. Second, a comparison of variables, analyses, and conclusions across studies is made to evaluate the divergence or convergence of research results. Chapter 6 points to some of the problems and pitfalls of using geographic information systems to examine environmental and public health issues. Chapter 7 uses the four resolutions (spatial, temporal, radiometric, and spectral) of remote sensing to analyze the contribution of satellite data in identifying and predicting risk areas for such diseases as leishmaniasis, trypanosomiasis (sleeping sickness), shistosomiasis, Rift Valley fever, malaria, hantavirus, Rocky Mountain Spotted Fever, Lyme disease, and onchocerciasis (river blindness). Chapter 8 discusses the specific processes of remote sensing and their ramifications for developing medical geography applications.

SYNOPSES OF THE INDIVIDUAL CHAPTERS

Chapter 2, "How Spatial Analysis Can be Used in Medical Geography," is a review of how geographers and others have used spatial analysis to study disease and health care delivery patterns. Point, line, area, and surface patterns, as well as map comparisons and relative spaces are discussed. Problems encountered in applying spatial analytic techniques are pointed out. The authors present some suggestions for the future use of spatial analytic techniques in medical geography.

Point pattern techniques include standard distance, standard deviational ellipses, gradient analysis and space and space-time clustering. Line methods include random walks, vectors and graph theory or network analysis. Under areas, location quotients, standardized mortality ratios, Poisson probabilities, space and space-time clustering, autocorrelation measures and hierarchical clustering are discussed. Surface techniques mentioned include isolines and trend surfaces. For map comparisons, coefficients of areal correspondence and correlation coefficients have been used. Case-control matching, acquaintance networks, multidimensional scaling and cluster analysis are examples of methods that are based on relative or non-metric space.

Chapter 2 continues with a discussion of several general points: problems encountered in spatial analysis, theory building and verification and the appropriate role of technique and computer use. Some suggestions are made for further use of spatial analytic techniques including more use of Monte Carlo simulation techniques, network analysis, environmental risk assessment, difference mapping, and multidimensional scaling.

Chapter 3, "Geographic Information Systems and Medical Geography," examines the use of geographic information systems to analyze spatial dimensions of health care services and disease distributions. This chapter chronicles the early years (through 1993) of the diffusion of geographic information systems into medical geography and related disciplines. It documents a small but vibrant body of research that was grappling with the introduction of GIS into the realms of health and disease. While some scholars were optimistically urging use of this emerging technology, others were advocating caution before jumping on the GIS bandwagon. All the while, a handful of investigators began to develop and operationalize applications of geographic information systems having specific foci on health and/or disease. Such applications as emergency response, AIDS prevention, hospital service areas, toxic air emissions, lead exposure, measles surveillance, radon risk, and cancer clusters are highlighted.

Chapter 4, "Geography Information Systems in Health Services Research," outlines research contributions that explore physician locations, hospital service and market areas, public health monitoring and surveillance programs, and emergency response planning within the context of geographic information systems. Aronoff's (1989) classification of GIS functions into (1) maintenance and analysis of the spatial data, (2) maintenance and analysis of the attribute data, (3) integrated analysis of the spatial and attribute data, and (4) cartographic output formatting functions provides a structure to evaluate the extent to which health services researchers have utilized the full potential of GIS. The chapter also presents multiple definitions of GIS and health services research, outlines some general concerns about geographic information systems, and makes a general appraisal of the contribution of this technology to the health of human populations.

Chapter 5, "GIS-Aided Environmental Research: Prospects and Pitfalls," is a fairly comprehensive review of the ways in which GIS can improve research into the human-environment relationship, as well as the special problems investigators encounter when they attempt to adapt this powerful analytic tool to such projects. The chapter catalogs the elements involved in human exposure from the toxicity of the pollutant through the ways the pollutant can change as it travels through the environment, to the final stage of manifesting in a diagnosable health effect. Two major groups of human-environment studies are being performed: analyses of the impact of existing hazards, and assessments of potential hazard from proposed industrial or residential developments in the planning phase.

Public health professionals will want to use this chapter as an aid in determining just how credible are their data, where they might go for additional data, and why combining data collected at different scales is risky. Not all statistical techniques are appropriate for studies such as these, either. Most of the commonly used techniques, such as analysis of variance and linear regression, assume that the observations were measured without error. These techniques are easily biased by characteristics common in the study of disease in space, such as the ways that events affect their surrounding areas and the ways that they influence future events in the same area. Techniques which are better able to handle these conditions without producing biased results are reviewed, such as mixed models, multilevel models, and structural equation modeling. Hopefully, the reader will find helpful suggestions for getting better results from ecologic studies that involve data collected at different scales, from the individual level to the aggregate.

Chapter 6, "Infectious Disease and GIS," reviews applications of geographic information systems that investigate spatial aspects of dracunculiasis (Guinea worm disease), LaCrosse encephalitis, Lyme disease, and malaria. For each infectious disease the text follows a sequence that includes a description of disease and its transmission chain, the geographic distribution and recent statistics, and a review of select research using geographic information systems. A cross-comparison of conclusions suggests that a targeted approach is more effective than broad-based approaches in eliminating or reducing vectors and corresponding rates of infection. These studies show the benefit of incorporating elements of human and physical geography into GIS databases used to combat vectored diseases.

Remote sensing is the process of collecting data about objects or landscape features without coming into direct physical contact with them. The application of remotely sensed data and image processing techniques can seem daunting and simply too expensive to implement. Chapters 7 and 8 are intended to take the novice remote sensing person through the entire process. Given the nature of this book, the focus is on the medical geography application of remotely sensed data. Chapter 7 is really the first part of a two-chapter sequence. It is intended that this chapter provide the framework to enable the layperson to act as an informed reader of the body of medical geography literature utilizing remotely sensed data. As such, it contains a brief history of remote sensing and introduces the basic vocabulary. The development of the technology of remote sensing parallels the use of the data within medical geography and helps to predict the direction of the discipline within the context of future applications.

Chapter 8 is a detailed look at the application of remotely sensed data within the existing body of medical geography literature. Each of the authors' use of the data is presented contextually in order to best explain the various techniques and to promote general comprehension, not only of the remote sensing vocabulary, but also in order to inspire ideas about how the data may be used in alternative case studies. Chapter 8 includes a number of technique-specific insets. These insets are designed to be more in-depth evaluations and discussions of the various methods used by the medical geography community when applying remotely sensed data. Chapter 8 also contains an overview of basic remote sensing terminology.

Both chapters may be reviewed independently, but of course are best understood within the context of the whole. These chapters intentionally differ from the existing body of medical remote sensing literature that usually follows a disease-specific formula in describing remote sensing applications. The approach used is application-specific rather than disease-specific in order to promote a more general understanding of the nature of the data and associated techniques applicable to a variety of diseases and disease vectors.

The chapters are interspersed with tables and figures that represent sample output from numerous geomedical applications of spatial analysis, GIS, and remote sensing applications. These tables and figures have been drawn from the original source articles with publishers' permissions. Instructors might use this volume as a source of illustrations useful in demonstrating geomedical applications of spatial analysis, GIS, and remote sensing.

This volume highlights geomedical applications of spatial analysis, geographic information systems, and remote sensing. Our aim is to describe "what" rather than "how." Knowing what has been done provides one with a sense of the big picture (i.e., current usage of geomedical GIS/RS applications). Knowing what also positions one to be able to springboard to extend existing applications or create new geomedical applications of spatial analysis, GIS, and remote sensing. Those requiring knowing how should consult the original source articles. To address how would require a detailed and technical account of data requirements and manipulations, software and hardware specifications, and the mathematics of geotechniques. This is beyond our scope since it is not the intent of this volume. Our reviews of particular geomedical applications highlight studies that build upon and extend one another. This seems a more rational approach than forcing the contents and findings of numerous and often redundant studies under a single subject heading (e.g., malaria, sleeping sickness, onchoceriasis). However, a master GIS/RS bibliographic reference guide includes some 400 articles that have been listed by subject.

This volume also includes a "Master GIS/RS Bibliographic Resource Guide," "Glossary," and "Index." The "Master GIS/RS Bibliographic Resource Guide" provides over 400 references to geomedical applications. Represented within this bibliography are citations from academic journals, trade publications, proceedings, and electronic documents (i.e., World Wide Web). The bibliography has been arranged by subject for the reader's convenience.

This volume also includes a glossary of spatial analysis, GIS, and remote sensing terminology. Here, terms from the text and other terms familiar to geoscientists are defined. To assist in accessing information, we have included both a subject and geographical index. We hope that combined, the appendix, bibliography, glossary, and indices constitute valuable reference tools for tapping the full potential of this resource guide as well as pointing to other outside sources.

A CAUTIONARY NOTE

The editors encourage readers to become grounded in the fundamental components and dynamics of their subject (health care system or disease) prior to forging on with geotechniques. It is important that one is knowledgeable about the basic sciences and/or clinical findings of the particular subject under investigation. Therefore, before diving headfirst into the realm of geomedical/technical application the following sequence is recommended.

- Know your subject. If you don't know, find out. It is very difficult to develop a sophisticated GIS application if you are not familiar with the health care service or disease under question. So, depending on your subject, you might want to become familiar with the organization, structure and dynamics of a health management organization; the factors influencing the prevention and transmission of diseases; the current spatial and temporal trends in disease incidence; and even the clinical symptoms of a particular disease.
- Read sections of this volume that relate to the subject area in which you are interested. If you need more information or more details, search the Master GIS/RS Bibliography to locate articles on your topic. Going to the original source often provides information as to the type of hardware, software, data, and analyses that were used in a particular study.
- Evaluate whether some of the existing GIS/RS applications highlighted in the text or referred to in the bibliography would be worth using or modifying for your project or program needs. Perhaps you have ideas that might enhance existing research. If your evaluation is affirmative, then...
- Explore the feasibility of developing your own GIS/RS capabilities (consult Aronoff, 1989), collaborating with existing GIS/RS facilities within your organization or system, or contracting out your project.
- Publish your results in official reports, newsletters, trade journals, and even academic journals so that others can benefit from your experience.

REFERENCES

Aronoff, S. 1989. *Geographic Information Systems: A Management Perspective*. Ottawa: WDL Publications.
Campbell, J.B. 1987. *Introduction to Remote Sensing*. New York: Guildford Press.

Dorland's Illustrated Medical Dictionary, 1985, 26th ed. Philadelphia: W.B. Saunders Company.

Environmental Systems Research Institute. 1996. *Introduction to ArcView GIS: Two-day Course Notebook with Exercises and Training Data.* Redlands, California: Environmental Systems Research, Inc.

European Conference on Environment and Health. 1990. *Environment and Health: The European Charter and Commentary: First European Conference on Environment and Health, Frankfurt, 7–8 December 1989.* Copenhagen: World Health Organization, Regional Office for Europe.

Hunter, J.M. 1974. The challenge of medical geography. In *The Geography of Health and Disease: Papers of the First Carolina Geographical Symposium,* J.M. Hunter (Ed.), pp. 1–31. Chapel Hill: University of North Carolina, Department of Geography.

Stedman, T.L. 1982. *Stedman's Medical Dictionary, Illustrated,* 24th ed. Baltimore: Williams and Wilkins.

White, K.L., J. Frenk, C. Ordonez, C. Paganini, and B. Starfield. 1992. *Health Services Research: An Anthology.* Washington, DC: Pan American Health Organization.

Chapter Two

How Spatial Analysis Can Be Used in Medical Geography

This chapter is an introduction to ways in which spatial analytic techniques can be used in the study of disease patterns and health care delivery, the two principal concerns of medical geography. A search through the literature on medical geography in the mid-1980s revealed that a great deal of interesting and useful work had used spatial analytic techniques as aids in understanding both disease patterns and health care delivery systems. The result was a review article (Gesler, 1986). Since that time, the literature has grown, most notably in two directions. First, some of the techniques described in the review article have become more sophisticated. Second, as predicted in the 1986 paper, GIS has been increasingly used in applying the techniques (Albert et al., 1995). Indeed, GIS technology has fostered a revival in the spatial analysis of health and disease phenomena, often facilitating the rapid calculation of appropriate formulas and the display of results. This chapter introduces the reader to a set of spatial analytic techniques that have and can be used by medical geographers and others. It also provides a useful bibliography of relevant research.

Why do we include this chapter in this book? For a start, medical geographers and others working in the health field should be aware that these kinds of studies exist. Others who work in the medical field expect that geographers will be acquainted with some basic applications of spatial analytic techniques. In addition, many situations arise where the appropriate technique would go a long way toward helping to solve a particular problem. The aware medical geographer should be in a position, perhaps with the aid of others more knowledgeable about spatial analysis, to select and apply the appropriate techniques.

Medical geographers will have differing opinions about what their field of study entails. The authors' boundaries for medical geography encompass: (1) the description of spatial patterns of mortality and morbidity, factors associated with these patterns, disease diffusion and disease etiology; (2) the spatial distribution, location, diffusion and regionalization of health care resources, access to and utilization of resources, and factors related to resource distribution and use; and (3) spatial aspects of the interactions between disease and health care delivery. This list of topics reflects the authors' knowledge and experience within medical geography. Therefore the studies reviewed here deal with these concerns. Other medical geographers might wish to include other topics. The material for this review was gathered from several of the leading geographic, epidemiological and social science journals and books published in North America and Britain. Undoubtedly, some important studies have been overlooked; one can only apologize for these omissions.

The first section of this paper presents findings from several medical studies that employed spatial analysis. This section is based on the dimensionality framework used by Unwin (1981) in his introductory book to spatial analysis. Thus points, lines, areas and surfaces will be discussed. This is, of course, a simplification; nevertheless, dimensionality aids in clarifying one's thinking. Besides the four types of dimensional study, map comparisons and relative spaces will also be considered. Within each type of research both descriptive and analytical techniques will be mentioned. Also, it will be noticed that applications to both disease and health care delivery studies are discussed under each dimensional heading. Table 2.1 summarizes the various methods medical geographers might find useful. The second part of the chapter addresses several points arising from the overview of the first part. Included here are discussions of problems inherent in spatial analysis, scale in particular; theory building and verification; the appropriate role of technique; and the use of computers. A final section makes some suggestions for future use of spatial analytic measures.

Unwin (1981) is a good starting point for those just becoming interested in this subject. Other recommended sources are Berry and Marble (1968), King (1969), Abler et al. (1971), Cliff et al. (1975), Unwin (1975), Ebdon (1977), Haggett et al. (1977), Tinkler (1977), Thomas (1979), Getis and Boots (1978), Journel and Huijbregts (1978), Kellerman (1981), Ripley (1981), Beaumont and Gatrell (1982), Diggle (1983), Gatrell (1983), Isaaks and Srivastava (1989), Cressie (1993), Haining (1990), and Bailey and Gatrell (1995). These books provide explanations of most of the techniques mentioned throughout this chapter (Table 2.1). Thus they can be used as guides for those unfamiliar with specific procedures. Also, the studies cited throughout the chapter often provide information on how techniques can be applied to particular problems.

The emphasis in this chapter is on techniques rather than study results. This means that in many cases examples of spatial analytic techniques might be taken out of the context of a piece of research for purposes of illustration. The dangers of this procedure are obvious, so interested readers are encouraged to follow up to see how a particular technique fits into an entire study. It can not be overemphasized that technique is only one part of the investigative process.

SPATIAL ANALYTIC TECHNIQUES

Point Patterns

There has been a great deal of interest in the analysis of point patterns of disease. From the start, we should distinguish between *general* methods which examine whether cases of a disease are clustered anywhere within a study area (looking for *clustering*) or *focused* methods which examine whether cases are clustered around a particular point of interest (looking for *clusters*). Unfortunately, it

Table 2.1. Spatial Analytic Techniques for Medical Geographers.

Points	**Surfaces**
Mean center/standard distance	Isolines
Standard deviational ellipse	Trend surface analysis
Gradient analysis	Power series polynomials
Nearest neighbor analysis	Fourier series
Variance/mean ratio test	
Quadrat analysis	**Map comparisons**
Space clustering	Lorenz curves
Space-time clustering	Coefficient of areal correspondence
	Correlation coefficient
Lines	Difference maps
Random walk	
Vectors	**Relative spaces**
Graph theory:	Case-control matching
Nodality	Acquaintance networks
Connectivity	Multidimensional scaling
Dispersion	Cluster analysis
Nodal hierarchies	
Flow analysis	
Areas	
Location quotients	
Standardized mortality ratios	
Poisson probability	
Space clustering	
Space-time clustering	
Autocorrelation measures	
Hierarchical clustering	

is not always clear whether researchers are investigating clustering or clusters. Dozens of methods have been devised to determine whether clustering or clusters are chance occurrences. Recently, GIS has come to the aid of clustering and cluster researchers. However, given an abundance of analytic techniques and new computer-aided technologies, there may be a tendency to ignore the processes underlying the spatial distributions of disease cases (Waller and Jacquez, 1995). That is, one should have an idea about the biological, environmental, or social mechanisms which might lead to various types of clustering or clusters. For example, one would expect noninfectious diseases such as certain types of cancers to be clustered around a hazardous waste site, while infectious diseases such as influenza might display a pattern of diffusion away from several nodes. It is important to distinguish between "true" and "perceived" clusters (Jacquez et al., 1996a). In true clusters, which explain fewer than five percent of all reported clusters, cases have a common etiology, whereas perceived clusters may arise due to chance or be made up of unrelated illnesses.

Researchers have discovered over the years that it is extremely difficult to "prove" that clustering or clusters have indeed occurred. Thus Wartenberg and Greenberg (1993) and others suggest that point pattern analysis should be undertaken to generate rather than test hypotheses. They "consider cluster studies to be pre-epidemiology: analytic investigations that can be done prior to more traditional, time-consuming and costly epidemiologic designs" (Wartenberg and Greenberg, 1993, p. 1764). They also emphasize the need for researchers to pay close attention to issues of statistical power and confounding. "Statistical power is the ability to detect an effect given that it is present" and "[C]onfounding is the erroneous attribution of an observation (or cluster) to a factor which is related to both an exposure (or risk factor) and an outcome (or disease)" (Wartenberg and Greenberg, 1993, p. 1764). Confounders include uneven population distributions, age, gender, ethnicity, and other factors.

Wartenberg and Greenberg (1993) set out four steps for the researcher to take when examining clusters. First, one has to *characterize the data*, which could be counts of disease events by geographical area, point locations of cases, event times, distances between events, counts of both cases and controls, and so on. Second, one must *decide the domain* from which the data come; this includes spatial, temporal, and space-time clusters. Third, one specifies a *null hypothesis* which is often that disease cases occur randomly. Fourth, one specifies an *alternative hypothesis*, typically that the distribution of cases deviates from a random pattern in a certain way, i.e., according to an underlying mechanism such as contagion or exposure to a contaminant.

As mentioned earlier, many methods are available for analyzing point patterns of disease occurrence. Early entrants into the field were nearest neighbor analysis and quadrat analysis. Pisani et al. (1984) used North's (1977) clustering method, which is based on the distance to nearest neighbor, to determine the degree of clustering among dwellings reporting variola minor (smallpox) in Braganca Paulista County, Brazil. The level of spatial clustering of cases was determined for different values of "defined distances" or fixed distances between dwellings with susceptibles and potential infective agents.

In his study of the diffusion of fowl pest disease in England and Wales, Gilg (1973) developed a frequency distribution based on outbreaks per grid square. From this quadrat analysis he calculated the mean/variance ratio to indicate whether the point pattern of outbreaks was clustered, random, or regular. The ratio also was an indication of what theoretical distribution might be fitted to the pattern. A form of quadrat analysis was part of Girt's (1972) examination of the relation of chronic bronchitis to urban structure in Leeds. He selected 30 quadrats and interviewed a sample of females in each quadrat. Comparison of his observed distribution of cases to the theoretical Poisson distribution showed significant variation among the quadrats. It should be noted that quadrat analysis is generally employed to assess overall point patterns for clustering, randomness or regularity. Here Girt identified particular quadrats that had more or fewer cases than expected by chance.

As Gatrell and Bailey (1996) point out, there is a basic flaw with nearest neighbor and quadrat techniques as used in human populations studies: they do not deal with the fact that people are not evenly distributed across space. Thus an apparent clustering or cluster of cases may simply be due to a clustering or cluster of people at risk; in other words, population distribution is a confounder. They suggest techniques that take this into account, such as comparing distributions of cases and controls taken from the population at large. Gatrell and Bailey also discuss techniques for exploring the first- and second-order properties of point patterns using a kernel estimation and K functions, taking as one example locations of childhood leukemia in west-central Lancashire.

Nearest neighbor and quadrat analysis techniques are restricted to one point in time. Of course, such processes as disease transmission take place over a period of time. If it can be shown that certain diseases occur in persons who are proximate in terms of certain combinations of distance and time, then perhaps contagion is indicated. This idea has given rise to a series of analytic techniques based on space-time clustering. Knox (1963) is given credit for the basic space-time clustering concept. He states that the detection of epidemicity in a set of data depends on a distribution in time, a distribution in space and interactions between these two dimensions. To examine interactions he asks whether pairs of cases which are relatively close in time are also relatively close in space. Pairs are classified according to both criteria and used to construct a contingency table. Observed pair frequencies can then be compared to expected values based upon a time interval distribution formula. Using this idea, Knox investigated the occurrence of cleft lip and palate among 574 children in Northumberland and Durham counties from 1949 to 1958. More recently, Knox and Gilman (1992) used more sophisticated space-time clustering techniques to examine leukemia clusters throughout Great Britain, and Knox (1994) compared leukemia clusters to specific map features, finding that there were associations between cases and railroads and fossil fuel-based hazards. As shall be shown later, space-time clustering has also been applied to areal data. A good source on space-time clustering can be found in Williams (1984).

Waller and Jacquez (1995) and Jacquez et al. (1996b) discuss several tests for both general and focused clustering, along with a table which sets out appropriate test statistics as well as null and alternative spatial models for each test. A few researchers have used a variety of "scan" or "moving window" techniques. Computer programs are written to move across a study area to detect areas where cases cluster. Gould et al. (1989) used this idea to examine suicides in the United States, Hjalmars et al. (1996) used it to detect clusters of childhood leukemia in Sweden, and Openshaw et al. (1987) developed a Geographical Analysis Machine (GAM) to look at leukemia clusters in northern England. Another method of recent origin is kriging, which is a smoothing or interpolation technique that "estimates the prevalence of a variable of interest at a given place using data from the surrounding regions" (Carrat and Valleron, 1992, p. 1293). Carrat and Valleron (1992) used kriging to map out an influenza-like illness epidemic in France, and Ribeiro et al. (1996) used the technique to examine the temporal and spatial distribution of anopheline mosquitoes in an Ethiopian village.

There has also been a limited amount of point pattern analysis in health care delivery studies; techniques used are generally much simpler than the methods we have just been discussing. As an example, using central place theory and concepts underlying the distribution of urban services as guides, Gober and Gordon (1980) investigated the location of physicians in Phoenix, Arizona. They compared their dot maps of locations to a four-celled model based on physician specialty and hospital orientation. Standard distance, the two-dimensional equivalent of the standard deviation, was used to determine relative clustering or dispersion among physician groups. This technique was also employed by Tanaka et al. (1981) to compare the changing patterns of population and health facility distribution

Table 2.2. Standard Distance of the Population and the Type of Clinical Function.

	1965	*1970*	*1975*
Population			
All ages	140	171	203
0–14 years	70	87	106
15–65 years	116	141	165
65 years and over	35	41	49
Clinical function			
Medicine	7.7	7.9	7.9
Pediatrics	8.2	9.3	9.3
Surgery	7.1	8.2	8.4
Obstetrics	8.3	8.2	8.4

Source: Social Science and Medicine, 15D, T. Tanaka, S. Ryu, M. Nishigaki, and M. Hashimoto. Methodological Approaches on Medical Care Planning from the Viewpoint of Geographical Allocation Model: A Case Study on South Tama District, pp. 83–91, 1981. Reprinted with permission from Elsevier Science.

in a Tokyo suburb between 1965 and 1975 (Table 2.2). Population potential was also used in this study to make similar comparisons.

The standard deviational ellipse provides more information than the standard distance measure as it also shows point pattern orientation and degree of eccentricity. The former descriptive measure was used by Shannon et al. (1978) to compare daily activity spaces and health-care-seeking spaces for black residents in Washington, DC; by Shannon and Cutchin (1994) to compare the distribution of population and general practitioners for different time periods in Munich, Germany (Table 2.3); and by Cromley and Shannon (1986) to map out activity spaces of elderly urban residents in Greater Flint, Michigan, and to relate these spaces to ambulatory medical care provision. Gesler and Meade (1988) used standard deviational ellipses to summarize daily activity patterns of respondents in a Savannah, Georgia, cardiovascular disease survey. Both the standard distance and standard deviational ellipses can provide information beyond the distribution of the point patterns they summarize. For example, they can provide clues to the influence of boundaries and transportation networks on activity patterns (Raine, 1978). A third descriptive point pattern technique, gradient analysis, was used by Giggs (1973) to investigate the distribution of schizophrenia in Nottingham. The proportions of 12 subgroups of patients who lived in a series of concentric rings around the city center were graphed to demonstrate the differential concentration of various types of patients.

Line Patterns

It seems that one-dimensional or line analysis has been used less for disease and health studies than the other dimensions in medical geography. One aspect of Brownlea's (1972) detailed investigation of the diffusion of infectious hepatitis

Table 2.3. Standard Deviational Ellipse Data, Munich General Practitioner Population.

Characteristic	1950	1960	1970	1980	1990
Mean center location					
x =	24.9	24.8	24.8	24.8	24.6
y =	8.7	8.7	8.7	8.6	8.6
Angle of orientation	89	88	84	84	81
Standard deviation in km					
x =	3.5	3.7	3.8	3.8	3.9
y =	4.2	4.2	4.6	4.8	4.8
Number of general practitioners	608	687	623	605	740

Source: Social Science and Medicine, 39, G.W. Shannon and M.P. Cutchin. General Practitioner Distribution and Population Dynamics: Munich, 1950–1990, pp. 23–38, 1994. Reprinted with permission from Elsevier Science.

in Wollongong, Australia, was the use of the concept of a random walk to analyze the movement of the disease's "clinical front." The idea here was to compare the actual direction of the disease movement with chance movements. Departures from expected directions would indicate that certain nonrandom constraints or "ecological parameters" might be at work in certain locations. Vectors or lines which indicate magnitude and direction can be used to describe or summarize disease movements and patient-to-health care resource flows. An example of the latter is Kane's (1975) vector displays of the health care-seeking behavior of residents of two rural counties in Utah.

Graph theory or network analysis has been used by medical geographers in both disease and health care delivery assessment. On the disease side, networks have been developed in diffusion studies to indicate various types of "joins" between the spatial units being investigated. These studies are really two-dimensional as they focus on join count measures among areal units. The networks themselves are convenient ways of depicting certain processes and are not analyzed in terms of such measures as connectivity or nodality. Thus Haggett (1976) developed seven alternative graphs to represent seven possible diffusion models of measles spread in southwestern England: regional, urban-rural, local-contagion, wave-contagion, journey-to-work, population size and population density. Adesina (1984) applied the same type of analysis to the spread of cholera in Ibadan in 1971. Brownlea's Wollongong study mentioned in the preceding paragraph used the network idea in a somewhat different manner. He first identified areas where annual disease notifications were outside the random (Poisson) range. The months in which these notifications were given were used to construct graphs which showed the changing origins and locations of the moving clinical front.

The work of Rogers (1979) demonstrates how graph theory can be applied to the diffusion of health care delivery systems. He traced the spread of family planning innovations among village women in South Korea using interpersonal relationships as the basic units of observation. Examination of the network elicited information on cliques, opinion leaders, connectivity, integration, diversity, openness and taboos. Harner and Slater (1980) attempted to regionalize hospitals in West Virginia by setting up a matrix of inter-county patient to hospital travel flows. Directed graphs were developed to analyze the flows (Figure 2.1). A directed line was defined to exist between a county population and a hospital if the probability of this flow was greater than selected fixed values varying from zero to one. Various fixed values or thresholds gave rise to a series of hierarchical clusters which aided in planning for better patient accessibility. Patient flows also lend themselves to interactive computer manipulation. Francis and Schneider (1984) reported on a graphics program which they used to map out referral patterns of cancer patients in western Washington State between 1974 and 1978 (Figure 2.2). They also provided several other examples of how their program could be used to help solve health care delivery problems. Probably the greatest use of graph theoretical concepts has been in the area of location/allocation modeling. The problem here is to locate a set of health care facilities and also to allocate sets of people to them in a way that produces some sort of optimal interaction between people and places. People and facilities can be represented as nodes and interactions or flows as weighted links. Abler et al. (1971) and Scott (1970) are good sources for overviews of the principal techniques that are involved. Godlund's (1961) use of location/allocation modeling to assign regional specialist hospitals for the government of Sweden is well known. Rushton (1975), among several others, has been very active in this area of medical geography.

Area Patterns

Maps of disease can be constructed in several ways. Some of the most common, like those that are based on natural breaks in rate distributions or the mean and standard deviations of distributions, are basically descriptive. Such methods as location quotients and standardized mortality ratios tend toward analysis and are generally more useful in pattern assessment. Many medical geographers have stressed the need for probability mapping, particularly for relatively rare diseases. There have been several instances in which the Poisson distribution has been used by medical geographers to identify units within a study area that have significantly high or low disease rates. White's (1972) investigation of leukemia in England and Wales is one example. Giggs et al. (1980) employed the Poisson probability test both to identify wards in Nottingham with high rates of primary acute pancreatitis and to show that the total number of cases and of female cases in one of Nottingham's six water supply areas was significantly greater than could have occurred by chance.

Gini indices, coefficients of localization, location quotients, and Lorenz curves are related and relatively simple, but informative, measures to assess inequalities in health care personnel and facility distributions. The Gini index and the coefficient of localization are statistics that gauge overall inequality across a study area, the Lorenz curve is a graphical display of inequality, and location quotients can be used to make choropleth maps showing where there are under- or over-supplies of resources.

All these methods are useful for comparing different study areas or changes in a study area over time. Readers can find the appropriate formulas and examples in Ricketts et al. (1994) and the articles reviewed here. Joseph and Hall (1985) calcu-

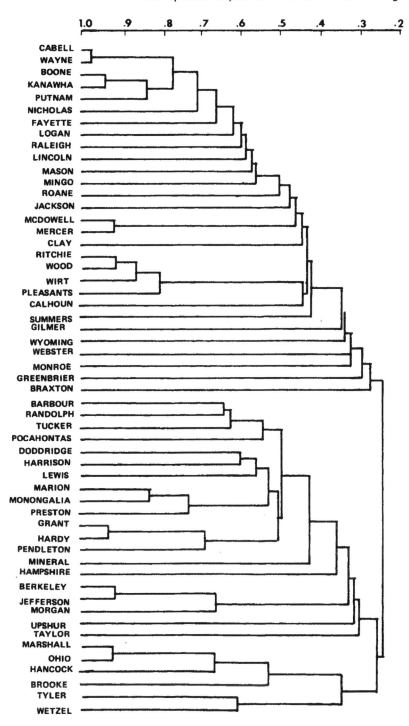

Figure 2.1. Dendogram Based on Inter-County and Intra-County Flows. *Source: Social Science and Medicine,* 14D, E.J. Harner and P.B. Slater. Identifying Medical Regions using Hierarchical Clustering, pp. 3–10, 1980. Reprinted with permission from Elsevier Science.

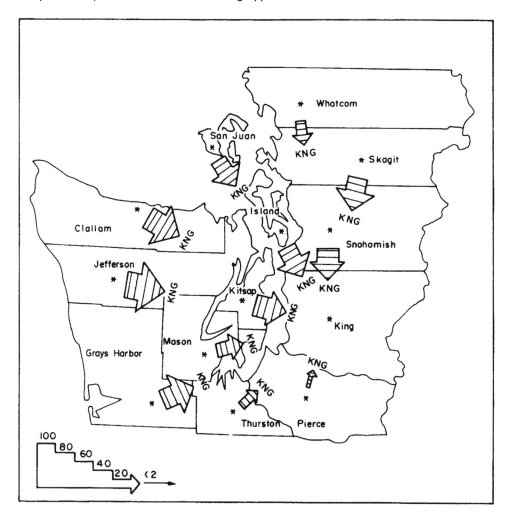

Figure 2.2. Percent of Cancer Patient Referrals to King County in 1974. *Source: Social Science and Medicine,* 18, A.M. Francis and J.B. Schneider. Using Computer Graphics to Map Origin-Destination Data Describing Health Care Delivery Systems, pp. 405–420, 1984. Reprinted with permission from Elsevier Science.

lated coefficients of localization and mapped out location quotients for three types of group homes (children's, adult, and psychiatric services) in Metropolitan Toronto and the City of Toronto. The Gini index was used by McConnel and Tobias (1986) to examine changes in the distribution of various types of physicians in the United States by states, counties, and SMSAs between 1963 and 1980. In their Munich study, Shannon and Cutchin (1994) used location quotients to map general practitioner locations in relation to population by district (Figure 2.3). Lowell-Smith (1993) used location quotients, Gini indices, and Lorenz curves in her examination of inequalities in the distribution of freestanding ambulatory surgery centers (FASCs) in the United States for the four major census regions, 48 states and District of Columbia, and metro and non-metro areas. Finally, Brown (1994) con-

Figure 2.3. General Practitioner Distribution and Population Dynamics. *Source: Social Science and Medicine,* 39, G.W. Shannon and M.P. Cutchin. General Practitioner Distribution and Population Dynamics: Munich, 1950–1990, pp. 23–38, 1994. Reprinted with permission from Elsevier Science.

ducted a very thorough analysis of the distribution of various types of health practitioners in Alberta using the Gini index, coefficient of localization, and Lorenz curve methods.

Tests for spatial clustering of disease, introduced in the section on point patterns, have also been developed for areal data. Ohno and Aoki (1981) devised a test procedure which they applied to three cancer mortality rates for 1123 city and county areas in Japan from 1969 to 1971. After classifying rates for each cancer into five categories, they identified all "concordant pairs": adjacent areal units whose rates fell into the same mortality category. A chi-square test was used to compare the observed concordant pairs with the expected number of such pairs.

As was the case with point pattern analysis, areal patterns have also been explored with space-time cluster methods. In the Nottingham study of primary acute pancreatitis by Giggs et al. (1980) mentioned above, Knox's method was used on 214 patients. They found no space-time clustering and concluded that the results did not support the hypothesis of an infective agent causing the disease. However, they suggested that there might be space-time clustering among patients in terms of workplace or previous residence.

Abramson et al. (1980) applied some basic techniques to look for both spatial and space-time clustering of Hodgkin's disease in Israel from 1960 to 1972. They uncovered 418 cases and matched these individually with controls who did not have the disease. Chi-square tests showed that cases and controls differed significantly in their geographic distribution over both the country's 14 administrative subdistricts and 40 "natural" regions. Giles (1983) also studied space-time clustering in Hodgkin's disease. To overcome latency and mobility problems, he suggested collecting historical data, particularly on residence and occupation. Since this information is usually not available for base populations, the case-control method is required. Armstrong (1976), who pioneered this type of study, compared the time spent in various Malaysian environments by nasopharyngeal cancer cases and controls.

Spatial autocorrelation analysis has been used in some interesting ways to examine disease patterns. Walter's (1993) paper introduces three indices of spatial autocorrelation (Moran's I, Geary's c, and a rank adjacency statistic D) and shows how they are affected by small sample size, by region, and by variations in age structure, in populations at risk, and in statistical power. Moran's I was used to look for autocorrelation in breast cancer mortality rates in Argentina (Wojdyla et al., 1996). Haggett's measles study, discussed in the preceding section on line patterns, used Moran's Black-White (BW) join count measure (free sampling) to examine the seven join graphs or models he had developed for contagion. Negative values of the standard normal deviate (z-score) of the test statistic showed a general tendency for spatial clustering or contagion, but z-scores varied considerably among the seven models. Haggett (1976) also compared diffusion patterns for the same graphs and for different graphs at different phases of the diffusion process. Finally, he speculated about how the models could be combined to provide a more accurate picture of measles spread. Adesina's (1984) work on cholera diffusion in Ibadan also used BW join counts to look for contagion; in this case five models and three phases of the process (advance, peak and retreat) were examined. Adesina also investigated the effects of different infection thresholds and tried to discover if there were directional biases in disease spread.

Glick (1979) has devised and tested several ways in which Moran's autocorrelation statistic for interval data can be used to examine spatial patterns of diseases and to look for biologic, chemical, physical, cultural and ethnic factors that might be associated with these patterns. The joins or weights used to calculate Moran's I statistic can be based on simple adjacency of geographical units, proportions of common boundaries, distance between the centers of the units, or whether two units fit into the same variable category (such as rural versus urban). In addition, spatial correlograms can be constructed which measure autocorrelation at different spatial lags (Figures 2.4 and 2.5). A lag of four, for example, indicates that units are "joined" only if there are three intervening units. Correlograms provide an indication of the scale at which spatial patterning is operating. Glick used these techniques to analyze sex-specific cancer mortality rates for nine body sites among the 67 counties of Pennsylvania (Table 2.4). In a study of skin cancer mortality in United States counties Glick (1982) went further with autocorrelation and other spatial analytic techniques. In this study he looked for trends in the autocorrelation function across linear transects and examined residuals from trend models. Lam et al. (1996) also made innovative use of correlograms. They examined the spread of AIDS in four regions of the United States (Northeast, California, Florida, and Louisiana) using county or parish data from 1982–1990 and were able to suggest when and where the spread was either mainly hierarchical or contagious.

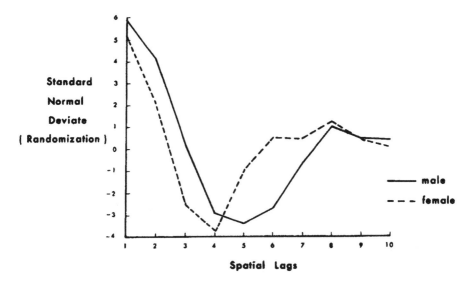

Figure 2.4. Spatial Correlation for Stomach Cancer. *Source: Social Science and Medicine,* 13D, B. Glick. The Spatial Autocorrelation of Cancer Mortality, pp. 123–130, 1979. Reprinted with permission from Elsevier Science.

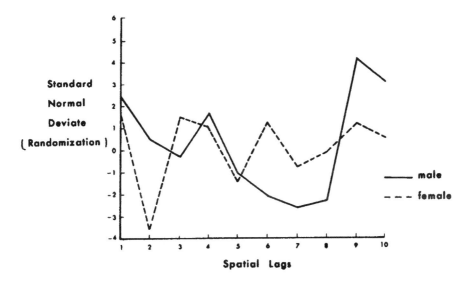

Figure 2.5. Spatial Correlation for Lung Cancer. *Source: Social Science and Medicine,* 13D, B. Glick. The Spatial Autocorrelation of Cancer Mortality, pp. 123–130, 1979. Reprinted with permission from Elsevier Science.

A method for determining hierarchical clusters of "high risk" areas has been developed by Grimson et al. (1981). The method commences by ranking disease rates for spatial units from high to low. The two highest ranking units are examined to see if they are adjacent or not, then adjacencies are counted among the highest three units, and so on. The observed number of adjacencies or joins are

Table 2.4. Spatial Autocorrelation Among First-Order Neighbors.

Cancer Type	I	Standard Normal Deviates	
		Randomization	Normalization
Stomach-m	0.43043	5.87766#	6.02389#
Stomach-f	0.37261	5.14608#	5.24220#
Lung-m	0.17159	2.47846*	2.52455*
Lung-f	0.11386	1.71151	1.74409
Breast-f	−0.03504	−0.26510	−0.26893
Cervix-f	0.08711	1.34898	1.38247
Bladder-m	0.19835	2.87729#	2.88639#
Leukemia-m	0.07954	1.26791	1.28018
Leukemia-f	−0.09869	−1.13502	−1.12932

* = Significant at 0.05 (two-tailed test).
# = Significant at 0.01 (two-tailed test).
Source: Social Science and Medicine, 13D, B. Glick. The Spatial Autocorrelation of Cancer Mortality, pp. 123–130, 1979. Reprinted with permission from Elsevier Science.

compared to the results of Monte Carlo, computer-simulated runs that Grimson performed in his analysis on cases of sudden infant death syndrome in the 100 counties of North Carolina. He found that significance was first reached for the eight highest ranking counties. There were also substantial increases in significance when 14, 18, and 24 counties were entered into the analysis.

Surface Patterns

A surface or scalar field can be constructed by using z or "height" values which correspond to x- and y-coordinates in two dimensions to draw isolines. Thus any disease or health care variables that have values for particular points in space can be mapped as a surface (Figure 2.6). Examples are Pyle and Lauer's (1975) maps of hospital market penetration areas based on proportions of spatial unit populations attending the hospital; Gilg's (1973) isoline maps using smoothed values of the date of first arrival, mode and mean by grid square for fowl pest disease diffusion; Mayhew's (1981) isochronal maps based on velocity fields drawn around emergency medical centers in large cities; Loytonen and Arbona's (1996) risk surface of obtaining HIV infection by municipality in Puerto Rico; and Rushton et al.'s (1996) contoured surface based on kriging of infant mortality and birth defect rates in the Des Moines, Iowa, urban region.

Two examples show how the well-known technique of trend surface analysis has been used to study diffusion processes; both involve power series polynomials. The first example is the study by Angulo et al. (1977) of variola minor spread in 1956 in Braganca Paulista County, Brazil. The following variables were used as the z-variable to develop linear, quadratic and cubic trend surfaces; time of the introduction of the disease into households for three types of introducers,

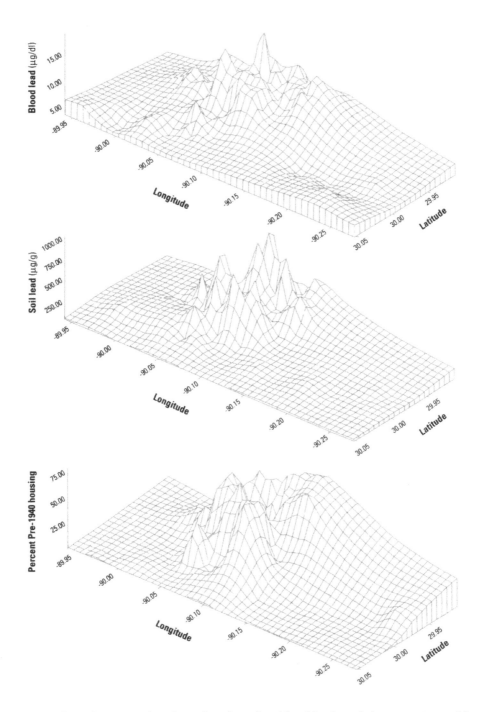

Figure 2.6. Three-dimensional surface plot of median blood lead, and the percentage of housing built before 1940 for New Orleans, Louisiana. z-Values were plotted on the same x- and y-coordinates of centroids for all census tracts with available New Orleans data. *Source: Environmental Health Perspectives,* 105, H.W. Mielke, D. Dugas, P.W. Mielke, K.S. Smith, S.L. Smith, and C.R. Gonzales. Associations Between Soil Lead and Childhood Blood Lead in Urban New Orleans and Rural Lafourche Parish of Louisiana, pp. 950–954, 1997. Reprinted with permission from NIH.

preschoolers, school children, and adults; school attended for school children; and several explanatory variables, including household size, number of susceptibles in a household, type of case and vaccination level. Kwofie (1976) applied trend surface analysis to the spread of cholera over a large area, West Africa, from 1970 to 1972. He also developed linear, quadratic and cubic surfaces and examined three periods (primacy, saturation and waning) of the diffusion process. Two major spatial trends, one along the coast, and one east-west across the interior Sahel, were uncovered.

MAP COMPARISONS

One technique available to geographers who wish to find out whether certain variables may help explain disease and health care resource patterns is map comparison. This can most easily be done by simply plotting dependent and independent variables and making visual comparisons. Thus, McGlashan (1972) conducted a survey in central Africa of 55 diseases and 20 environmental factors that might have been associated with the diseases. Data came from patient records at 84 hospitals. Visual examination of disease and factor maps led McGlashan to carry out some contingency table analyses. For example, he compared the number of annual cases of *diabetes mellitus* with whether cassava was the staple food eaten by hospital patients.

Probably the most used statistical method of map comparison is correlation analysis or "ecological correlation." Here health care resource or disease rates for spatial units are compared using Pearson's product-moment or Spearman's rank correlation statistics. Pyle (1973) found no strong correlations when he compared census tract maps of measles incidence in Akron, Ohio, for 1970–1971 with maps of various demographic and socioeconomic variables. However, when he performed a hierarchical clustering technique (based on 12 census variables) on the tracts to form five regions, the two poverty areas did correspond with concentrations of measles cases. Ecological correlation was also used by Gesler et al. (1980) to compare maps of community characteristics to disease reporting and hospital use by census tract in Central Harlem Health District, New York City. Both individual variables and factor scores from a factor analysis of community variables were correlated with the dependent variables. Most of these aggregate findings corresponded to results of studies of individuals. A third example of ecological correlation comes from Smith's (1983) study of the geographic distribution of alcohol treatment facilities in Oklahoma. In this investigation Smith correlated an index of service comprehensiveness with need, urbanization, income and attitudes toward alcohol use by county.

Another type of map comparison technique that does not appear to have been used much by medical geographers is based on the coefficient of areal correspondence, which is the ratio of the area over which two phenomena are located together to the total area covered by the two phenomena. Court's (1970) modification of this technique for surfaces was used by Hugg (1979) to compare the geographic distribution of work disability and poverty status for persons 18 to 64 using the 50 states of the United States as units of observation.

BEYOND SPATIAL ARRANGEMENTS

Gatrell (1983) calls the dimensional analyses which have just been discussed the study of spatial arrangements. They are based on absolute space and the metric properties of distance. This, Gatrell suggests, is just the beginning of spatial analysis. Geographers need to go beyond spatial arrangement to consider relative spaces and nonmetric relationships among sets of objects. The following examples show some innovative ways in which the concepts of "space" and "distance" have been used in medical research.

In an analysis of factors related to cardiovascular deaths in Evans County, Georgia, Smith et al. (1977) tackled the problem of the best way to match a small set of cases with one or more controls that possessed the "same" values for certain variables. Categorical variables like sex and race required an exact match. For continuous variables like age and systolic blood pressure, a minimum "distance" was calculated. This distance was the sum of the differences between the z-scores (based on case variable distributions) of cases and controls for all continuous variables. The "nearest" control was selected as a match.

Greenwald et al. (1979) examined a transmissibility or clustering hypothesis for the relatively rare diseases of leukemia and lymphoma by developing acquaintanceship networks among case and control pairs. Twenty lymphoma and 17 leukemia cases were found in Orleans County, New York, for the period 1967–1972. Data were gathered on acquaintances and acquaintances of acquaintances for the 37 cases and also 37 controls; in all 13,409 people were involved. Four types of pairs were possible, case-case, case-control, control-case and control-control. The analysis focused on pairs with two or more intermediate links. The null hypothesis was based on a permutation distribution. The researchers stated that their method attempted to avoid the problems of space-time clustering techniques: namely, long latency period and reliance on the date of diagnosis to establish disease onset.

Multidimensional scaling promises to be a useful tool for medical geographers. Ninety students at the University of Oklahoma were asked by Smith and Hanham (1981) to evaluate 28 public facilities on "noxiousness." The INDSCAL algorithm was applied to similarity matrices of responses. Three dimensions proved to be of importance, noxious/desirable, physical services/human services and residential/treatment. Mental health facilities, as expected, were seen as especially noxious. In an earlier paper, Dear et al. (1977) also reported on reactions to mental health care facilities, in this case community reaction to their location. Multidimensional scaling was used to identify important attributes by which people judged ten mental health facilities in Philadelphia.

DISCUSSION

Problems in Spatial Analysis

This review of the ways in which medical geographers and others have applied spatial analytic techniques to disease and health care situations has led to several generalizations. The first of these is that there are certain problems associated with the use of these techniques. Some of the problems are general and some specific to particular methods of analysis. A good place to begin a study of these problems is Unwin's introduction to spatial analysis (1981). Mayer (1983) has written about the epistemological, logical, and methodological problems that one faces in spatial analyses that attempt to detect disease causation or etiology. Another good source is King (1979) who discussed the following difficulties that arise in geographical epidemiology: the necessity for aggregating disease rates over space and time which gains data stability but loses information; accuracy of death certificates and diagnoses; choice of a suitable rate standardization procedure; choice of scale and data classes when constructing maps of disease rates; modifiable units; and ecologic fallacies. Stimson (1983) has pointed out several pitfalls in conducting studies on health care delivery. These include inaccuracies, incompleteness and instability of data sources; making unwarranted causal inferences from ecological data; using data that are not disaggregated to the smallest level of scale possible; and comparing data sets that do not correspond in scale and time. Most of these problems are familiar to geographers, but should nonetheless always be kept in mind.

It is always refreshing to find researchers admitting that their particular study has encountered difficulties. An example is Haggett's (1976) measles diffusion work for which he reports the problems of unit aggregation, cross unit flows, size and population differences among the units of observation, and unit linkage definitions. McGlashan (1972) has acknowledged the frustration that many of those studying developing countries have with the lack of, and inaccuracy of, data. Generally speaking, analyses based on data from these countries cannot be very sophisticated. Sometimes investigators attempt to circumvent such problems with a new methodology. Greenwald et al.'s (1979) use of acquaintanceship networks to examine rare diseases (above) is an example of overcoming a data problem with a new method.

The problem of scale is of course simply part of being a geographer. Medical geographers have often pointed out that spatial patterns or variable associations show striking differences at different scales of analysis. Because of this phenomenon, disease and health care investigations should be carried out at several different geographic levels. Schneider et al. (1993) make this point when they show that evidence for cancer clusters varies a great deal at the four scales they used in a New Jersey study: state level, degree of urbanization, counties, and minor civil divisions. In a study of the relationship between infant mortality and birth defect rates in Des Moines, Rushton et al. (1996) found that spatial patterns were sensitive to the size of the spatial filters (0.4 miles and 0.8 miles). Waller and Turnbull (1993) demonstrate that the performance of three statistical tests used to detect the presence of clusters of adverse health effects is sensitive to the level of aggregation of data.

It can be argued that scale should be used creatively (Cleek, 1979). For one thing, replication of findings at different scales tends to confirm hypotheses. Scale can also be used to synthesize a set of investigations on the same topic which have been carved out for different population levels and data aggregation. Along these lines, Meade (1983) investigated cardiovascular disease at different scales both within the Enigma Area of the southeastern United States and within the city of Savannah, Georgia. Furthermore, as White (1972) says, one can try to identify the scale at which a certain process is most effective; this in itself may provide clues as to how the process works. This is one of the ideas behind the use of spatial correlograms. Scale has a particularly interesting part to play in diffusion analyses. Angulo et al. (1979) demonstrate how different types of diffusion (hierarchical, contagious, etc.) operate at different levels of data and population aggregation.

Theory Building and Verification

If, as Mayer (1983) states, geographic patterns arise from underlying processes, then medical geographers equipped to theorize about processes propelling pathogen transport or clinic location will make the best use of spatial analytic techniques. In other words, theory must be a part of studies that use spatial analytic techniques. In some cases spatial analysis aids in theory building or provides clues to the underlying processes. For example, standard deviational ellipses of health care movements may suggest that certain boundaries affect behavior. If sex-specific maps of cancer mortality rates are similar in pattern, then environmental factors may be implicated; if they are different, then one might look more closely at occupation or behavior. In contrast to theory-generating studies are those that begin with a theory and attempt to confirm or reject it. Usually, studies of disease clusters in space or in time-space test a contagion or transmissibility hypothesis. Strategies for investigating urban physician location patterns may be based on ideas about urban ecological structure or on central place notions like hierarchies, thresholds and ranges.

Several geographers have pointed out the major difficulties of connecting spatial patterns and processes: some processes can generate many spatial patterns, and the same pattern may result from many different processes. The former problem arises because processes are stochastic or give rise to chance variations. The latter problem indicates the need for *a priori* knowledge about the situation so that the appropriate process(es) will be studied. A good example of this, applicable to disease spread, is the difficulty of distinguishing between generalized and compound point patterns. In a generalized or true contagion process the first points are randomly located and then others cluster around these. In the compound or apparent contagion case the distribution of points is related to some other phenomenon which is unevenly distributed across the space. In the latter process clusters are also formed, but this is not due to contagion.

Medical geographers who are alert to both spatial and medical matters are most likely to produce genuinely useful findings. Because they deal with a multitude of factors outside, but related to, geography and health, they must blend theories from different disciplines. In the area of disease causation, Mayer (1983) has pointed out that epidemiologists use few spatial techniques and geographers know little about pathogenesis or biological processes. Other disciplines, and the other social sciences in particular, have contributed much to the study of health care delivery, and medical geographers can exploit this knowledge base to good effect.

The studies reviewed here reveal the vitality of geographic thought as it contributes to theories about health and disease. This vitality stems from the training of geographers in urban, economic, physical, political, environmental and cultural geography. There are many instances where geographic research has contributed to changing existing theories or formulating new ones; spatial analysis has the potential to help further this tradition.

The Appropriate Role of Technique

A review of this type, where certain spatial analytic techniques have been extracted from research reports, could easily give the impression that technique is all. The studies cited here, however, are proof that few medical geographers would make this mistake. There are at least three indications of an awareness of the proper role of technique: (1) The use of several different techniques within a single investigation. While this seems to lay undue emphasis on technique it also shows that there can be flexibility in trying out different methods to solve different aspects of a problem. The work of Giggs et al. (1980), Giles (1983), Gilg (1973), Girt (1972), and Glick (1982) are all good examples of this point. (2) The reluctance of most geographers to say that quantification alone is the complete answer. This is the recognition that theory, process, description and explanation are just as important as analytic methods. (3) An awareness that spatial analysis alone is not sufficient. Angulo et al. (1979) illustrate this when they report that some diffusion links do not depend on geographic proximity only, but also on social proximity such as which school a child attends. The section on going beyond spatial arrangements also points in this direction.

Computer Use

When Gesler (1986) first wrote about the uses of spatial analysis in medical geography over a dozen years ago, he made an easy prediction that the techniques discussed would become more computerized. The computer revolution that was beginning then is now being realized. Large data sets and complicated algorithms can now be handled relatively quickly and efficiently by modern machines. This of course opens up opportunities for medical geographers, provided the data is there and the technique used is valid. Researchers have a variety of GIS and statistical software with which they can visualize, explore, and model disease and health data sets (Gatrell and Bailey, 1997). As examples, Rushton and Lolonis (1996) used the address-matching capabilities of a GIS and TIGER files to map birth defects in Des Moines, Hjalmars et al. (1996) employed ARC/INFO software to search for childhood leukemia clusters in Sweden, and Waller and Jacquez (1995) suggested that a GIS be used as a proactive disease surveillance strategy. One of the more well-known applications of a GIS is that by Openshaw et al.'s (1987) geographical analysis machine ("GAM," in the Point Pattern section above). The interactive use of computers to study such things as patient flows is also of potentially great value (Walsh et al., 1997). Moreover, as location/allocation algorithms become more complex their dependence on computers increases.

Computer programs are especially valuable in various simulation studies establishing probabilities for join occurrence among the appropriate number of areal units. Monte Carlo simulation is recommended by Grimson et al. (1981) for situations where spatial data are not independent and areal units are of irregular shape. It may not be statistically correct to compare observed and theoretical patterns under these conditions. Monte Carlo techniques may also be required if no theoretical distributions have been formulated. An example is provided by Ohno and Aoki (1981). To check the validity of their chi-square results they simulated a theoretical distribution of adjacent concordant pairs and determined from this distribution the probability that their observed number of pairs could occur by chance. For the most part, results confirmed their other findings. Using data on a smallpox epidemic in nineteenth century Finland, Wilson (1993) developed "a simulation model to illustrate why observed smallpox mortality patterns are an incomplete picture of their underlying morbidity patterns" (p. 277). Loytonen and Arbona (1996) used a Monte Carlo simulation technique to forecast the spatial diffusion of AIDS in Puerto Rico. The significance of findings can also be determined using simulation methods. One example is the use of simulation to find out if the spatial distribution of infant mortality and birth defects in Des Moines was due to chance alone (Rushton and Lolonis, 1996; Rushton et al., 1996). Another is Gatrell and Bailey's (1996) construction of simulation envelopes to graphically assess the evidence for clustering in childhood leukemia in west-central Lancashire, 1954–1992.

An important introductory article for exploring the use of GIS and complementary statistical software for spatial analysis is by Gatrell and Bailey (1997), who focus on point patterns. They discuss the techniques that can be performed using several proprietary GIS packages, including IDRISI, GRASS, GRID, and ARC/INFO. They also show how spatial analysis packages such as INFOMAP and S+ can be used to examine health data and demonstrate how SPLANCS, a spatial analysis system built around S+ for handling point data, can be used. Finally, they discuss the "close coupling" of S+ and ARC/INFO which enables exchanges between the statistical and GIS programs. Chapter 7 of the *S+ Spatial Stats: User's Manual for Windows and UNIX* (Kaluzny et al., 1998) is a practical guide to coupling S+ with ARC/INFO.

FURTHER STUDY

Any suggestions for the future use of spatial analytic techniques in medical geography must be made with great hesitancy as their feasibility can only be guessed. A few ideas that came to mind during the preparation of this review will be noted here simply as examples of what could be done. (1) More use might be made of Monte Carlo simulation techniques in assessing point patterns of disease. If the population which contains disease cases is not randomly distributed over space, then it is not statistically valid to compare the distribution of cases with known theoretical distributions like the Poisson, Neyman type A or negative binomial. Monte Carlo methods can be used to determine whether the pattern of cases is more clustered than the population itself. Some interesting problems, such as whether individuals or households are the units of observation, will be encountered here. (2) Network analysis might be used to examine patient referrals. Referral systems seem to be little understood. Patient origins and the locations of physicians and health care facilities could be graphed as network nodes and patient flows could be graphed as weighted links. The usual tests of nodality, connectivity, hierarchical linkages and so on, could be applied. (3) Network analysis might also be applied to the study of health care delivery to pastoralists. The problems of trying to service nomadic peoples are well known. Perhaps points where they tend to stay for relatively long periods during their migratory cycles could be used as network nodes. Analysis might reveal key nodes for service contact. (4) Geographic information systems can be used to assess environmental risk. The ability to overlay disease patterns and possible causative factor patterns on the same x- and y-coordinate system should prove to be fruitful in the future. (5) Difference mapping, a method based on join counts, could be useful for comparing disease maps with maps of factors that might be associated with disease or with maps of theoretical disease patterns (Cliff, 1970). The method is based on join counts. Whereas other map comparison techniques like correlation and coefficients of areal correspondence depend on a single number, difference maps consider the spatial arrangement of map similarities. (6) Multidimensional scaling might be used to assess social distance between patients and health care providers. The problem of social distance as a barrier to health care has been recognized in cultures around the world. The difficulty is how to measure this seemingly subjective variable. No doubt sociologists and anthropologists have addressed this question, but medical geographers can also address it alongside other "distance" measures.

CONCLUDING REMARKS

Applying spatial analysis in medical geographic research can be tedious and sometimes disappointing. First, one must select a technique that is appropriate to the problem at hand. Next, all the problems discussed above must be tackled. Then comes the difficulty of understanding the technique itself. Many of those who write about spatial analysis make little effort to make their procedures clear to the mathematically unsophisticated reader. The next step is to operationalize techniques, which may require some relatively advanced knowledge of computer programming. Finally, once the technique has been applied, its usefulness in analysis or explanation should be honestly appraised. If it is inappropriate, then it should be discarded.

Undoubtedly, there are many medical geographers who are reluctant to use spatial analytic techniques, however appropriate they might appear to be, because of the mathematical and computer programming skills required to understand and operationalize them. Most geographers do not have the necessary background; thus there is a gap between the potential and the actual use of spatial analysis to help solve interesting problems. A major aim of this chapter is to acquaint the reader with potential uses and thus make technique more accessible. Utilization of technique, however, can only come with the study of specific procedures and how they have been employed in research. Many medical geographers will probably want to collaborate with others possessing more knowledge of analytic techniques; such collaborations have produced several recent papers in medical geography.

REFERENCES

Abler, R., J.S. Adams, and P. Gould. 1971. *Spatial Organization: The Geographer's View of the World*. Englewood Cliffs, NJ: Prentice-Hall.

Abramson, J.H., N. Goldblum, M. Avitzur, H. Pridan, M.I. Sacks, and E. Peritz. 1980. Clustering of Hodgkin's disease in Israel: A case-control study. *International Journal of Epidemiology* 9(2):137–144.

Adesina, H.O. 1984. Identification of the cholera diffusion process in Ibadan, 1971. *Social Science and Medicine* 18(5):429–440.

Albert, D.P, W.M. Gesler, and P.S. Wittie. 1995. Geographic information systems and health: An educational resource. *Journal of Geography* 94(2):350–356.

Angulo, J.J., C.K. Takiguti, C.A. Pederneiras, A.M. Carvalho-de-Souza, M.C. Oliveira-de-Souza, and P. Megale. 1979. Identification of pattern and process in the spread of a contagious disease. *Social Science and Medicine* 13D:183–189.

Angulo, J.J., P. Haggett, P. Megale, and A.A. Pederneiras. 1977. Variola minor in Braganca Paulista County, 1956: A trend-surface analysis. *American Journal of Epidemiology* 105(3):272–278.

Armstrong, R.W. 1976. The geography of specific environments of patients and non-patients in cancer studies, with a Malaysian example. *Economic Geography* 52:161–170.

Bailey, T.C. and A.C. Gatrell. 1995. *Interactive Spatial Data Analysis*. New York: Wiley.

Beaumont, J.R. and A.C. Gatrell. 1982. *An Introduction to Q-Analysis*. CATMOG 34. Norwich: Geo Abstracts.

Berry, B.J.L. and D.F. Marble. 1968. *Spatial Analysis: A Reader in Statistical Geography*. Englewood Cliffs, NJ: Prentice-Hall.

Brown, M.C. 1994. Using Gini-style indices to evaluate the spatial patterns of health practitioners: Theoretical considerations and an application based on Alberta data. *Social Science and Medicine* 38(9):1243–1256.

Brownlea, A.A. 1972. Modelling the geographic epidemiology of infectious hepatitis. In *Medical Geography: Techniques and Field Studies*, N.D. McGlashan (Ed.), pp. 279–300. London: Methuen.

Carrat, F. and A.-J. Valleron. 1992. Epidemiologic mapping using "kriging" method: Application to an influenza-like illness epidemic in France. *American Journal of Epidemiology* 135(11):1293–1300.

Cleek, R.K. 1979. Cancers and the environment: The effect of scale. *Social Science and Medicine* 13D(4):241–247.

Cliff, A.D. 1970. Computing the spatial correspondence between geographical patterns. *Transactions of the Institute of British Geographers* 50:143–154.

Cliff, A.D., P. Haggett, J. Ord, K. Bassett, and R. Davies. 1975. *Elements of Spatial Structure: A Quantitative Approach*. Cambridge, MA: Cambridge University Press.

Court, A. 1970. Map comparisons. *Economic Geography* 46:435–438.

Cressie, N.A.C. 1993. *Statistics for Spatial Data*. New York: Wiley.

Cromley, E.K. and G.W. Shannon. 1986. Locating ambulatory medical care facilities for the elderly. *Health Services Research* 21(4):499–514.

Dear, M., R.F. Fincher, and L. Currie. 1977. Measuring the external effects of public programs. *Environment and Planning A* 9:137–147.

Diggle, P.J. 1983. *Statistical Analysis of Spatial Point Patterns*. London: Academic Press.

Ebdon, D. 1977. *Statistics in Geography: A Practical Approach*. Oxford: Blackwell.

Francis, A.M. and J.B. Schneider. 1984. Using computer graphics to map origin-destination data describing health care delivery systems. *Social Science and Medicine* 18(5):405–420.

Gatrell, A.C. 1983. *Distance and Space: A Geographical Perspective*. Oxford: Clarendon Press.

Gatrell, A.C. and T.C. Bailey. 1996. Interactive spatial data analysis in medical geography. *Social Science and Medicine* 42(6):843–855.

Gatrell, A.C. and T.C. Bailey. 1997. Can GIS be made to sing and dance to an epidemiological tune? In *Proceedings of the International Symposium on Computer Mapping in Epidemiology and Environmental Health*, R.T. Aangeenbrug, P.E. Leaverton, T.J. Mason, and G.A. Tobin (Eds.), pp. 38–52. Alexandria, VA: World Computer Graphics Foundation.

Gesler, W. 1986. The uses of spatial analysis in medical geography: A review. *Social Science and Medicine* 23(10):963–973.

Gesler, W.M. and M.S. Meade. 1988. Locational and population factors in health care-seeking behavior in Savannah, Georgia. *Health Services Research* 23(3):443–462.

Gesler, W.M., C. Todd, C. Evans, G. Casella, J. Pittman, and H. Andrews. 1980. Spatial variations in morbidity and their relationship with community characteristics in Central Harlem Health District. *Social Science and Medicine* 14D(4):387–396.

Getis, A. and B. Boots. 1978. *Models of Spatial Processes: An Approach to the Study of Point, Line, and Area Patterns*. Cambridge, MA: Cambridge University Press.

Giggs, J.A. 1973. The distribution of schizophrenia in Nottingham. *Transactions of the Institute of British Geographers* 59:55–76.

Giggs, J.A., D.S. Ebdon, and J.B. Bourke. 1980. The epidemiology of primary acute pancreatitis in the Nottingham Defined Population Area. *Transactions of the Institute of British Geographers* 5:229–242.

Giles, G.G. 1983. The utility of the relative risk ratio in geographic epidemiology: Hodgkin's disease in Tasmania 1972–1980. In *Geographical Aspects of Health: Essays in Honor of Andrew Learmonth*, N.D. McGlashan and J.R. Blunden (Eds.), pp. 361–374. London: Academic Press.

Gilg, A.W. 1973. A study in agricultural disease diffusion: The case of the 1970–71 fowl-pest disease. *Transactions of the Institute of British Geographers* 59:77–97.

Girt, J.L. 1972. Simple chronic bronchitis and urban ecological structure. In *Medical Geography: Techniques and Field Studies*, N.D. McGlashan (Ed.), pp. 211–230. London: Methuen.

Glick, B. 1979. The spatial autocorrelation of cancer mortality. *Social Science and Medicine* 13D(2):123–130.

Glick, B.J. 1982. The spatial organization of cancer mortality. *Annals of the Association of American Geographers* 72(4):471–481.

Gober, P. and R.J. Gordon. 1980. Intraurban physician location: A case study of Phoenix. *Social Science and Medicine* 14D(4):407–417.

Godlund, S. 1961. Population, regional hospitals, transportation facilities, and regions: Planning the location of regional hospitals in Sweden. In *Lund Studies in Geography No. 21, Series B.* Lund: Department of Geography, Royal University of Lund.

Gould, M.S., S. Wallenstein, and L. Davidson. 1989. Suicide clusters: A critical review. *Suicide and Life-Threatening Behavior* 19(11):17–29.

Greenwald, P., J.S. Rose, and P.B. Daitch. 1979. Acquaintance networks among leukemia and lymphoma patients. *American Journal of Epidemiology* 110(2):162–177.

Grimson, R.G., K.C. Wang, and P.W.C. Johnson. 1981. Searching for hierarchical clusters of disease: Spatial patterns of sudden infant death syndrome. *Social Science and Medicine* 15D(2):287–293.

Haggett, P. 1976. Hybridizing alternative models of an epidemic diffusion process. *Economic Geography* 52:136–146.

Haggett, P., A.D. Cliff, and A. Frey. 1977. *Locational Methods.* London: Arnold.

Haining, R.P. 1990. *Spatial Data Analysis in the Social and Environmental Sciences.* Cambridge, MA: Cambridge University Press.

Harner, E.J. and P.B. Slater. 1980. Identifying model regions using hierarchical clustering. *Social Science and Medicine* 14D:3–10.

Hjalmars, U., M. Kulldorff, G. Gustafsson, and N. Nagarwalla. 1996. Childhood leukaemia in Sweden: Using GIS and a spatial scan statistic for cluster detection. *Statistics in Medicine* 15(7/9):707–715.

Hugg, L. 1979. A map comparison of work disability and poverty status in the United States. *Social Science and Medicine* 13D(4):237–240.

Isaaks, E.H. and R.M. Srivastava. 1989. *Applied Geostatistics.* New York: Oxford University Press.

Jacquez, G.M., L.A. Waller, R. Grimson, and D. Wartenburg. 1996a. The analysis of disease clusters, Part I: State of the art. *Infection Control and Hospital Epidemiology* 17(5):319–327.

Jacquez, G.M., R. Grimson, L.A. Waller, and D. Wartenburg. 1996b. The analysis of disease clusters, Part II: Introduction to techniques. *Infection Control and Hospital Epidemiology* 17(6):385–397.

Joseph, A.E. and G.B. Hall. 1985. The locational concentration of group homes in Toronto. *Professional Geographer* 37(2):143–154.

Journel, A.G. and C.J. Huijbregts. 1978. *Mining Geostatistics.* London: Academic Press.

Kaluzny, S.P. et al. 1998. S+ *Spatial Stats: User's Manual for Windows and UNIX.* New York: Springer.

Kane, R.L. 1975. Vector resolution: A new tool in health planning. *Medical Care* 13(2):126–136.

Kellerman, A. 1981. *Centrographic Measures in Geography.* CATMOG 32. Norwich: Geo Abstracts.

King, L.J. 1969. *Statistical Analysis in Geography.* Englewood Cliffs, NJ: Prentice-Hall.

King, P.E. 1979. Problems of spatial analysis in geographical epidemiology. *Social Science and Medicine* 13D(4):249–252.

Knox, E.G. 1994. Leukaemia clusters in childhood: Geographical analysis in Britain. *Journal of Epidemiology and Community Health* 48(4):369–376.

Knox, E.G., and E. Gilman. 1992. Leukaemia clusters in Great Britain. 2. Geographical concentrations. *Journal of Epidemiology and Community Health* 46(6):573–576.

Knox, G. 1963. Detection of low intensity epidemicity: Application to cleft lip and palate. *British Journal of Preventive and Social Medicine* 17:121–127.

Kwofie, K.M. 1976. A spatio-temporal analysis of cholera diffusion in Western Africa. *Economic Geography* 52:127–135.

Lam, N.S., M. Fan, and K. Liu. 1996. Spatial-temporal spread of the AIDS epidemic, 1982–1990: A correlogram analysis of four regions of the United States. *Geographical Analysis* 28(2):93–107.

Lowell-Smith, E.G. 1993. Regional and intrametropolitan differences in the location of freestanding ambulatory surgery centers. *Professional Geographer* 45(4):398–407.

Loytonen, M. and S.I. Arbona. 1996. Forecasting the AIDS epidemic in Puerto Rico. *Social Science and Medicine* 42(7):997–1010.

Mayer, J.D. 1983. The role of spatial analysis and geographic data in the detection of disease causation. *Social Science and Medicine* 17(16):1213–1221.

Mayhew, L.D. 1981. Automated isochrones and the location of emergency medical services in cities. *Professional Geographer* 33:423–428.

McConnel, C.E. and L.A. Tobias. 1986. Distributional change in physician manpower, United States, 1963–80. *American Journal of Public Health* 76(6):638–642.

McGlashan, N.D. 1972. Geographical evidence on medical hypotheses. In *Medical Geography: Techniques and Field Studies,* N.D. McGlashan (Ed.). London: Methuen.

Meade, M. 1983. Cardiovascular disease in Savannah, Georgia. In *Geographical Aspects of Health: Essays in Honour of Andrew Learmonth,* N.D. McGlashan and J.R. Blunden (Eds.), pp. 175–196. London: Academic Press.

North, P.M. 1977. A novel clustering method for estimating numbers of bird territories. *Applied Statistics* 26:149.

Ohno, Y. and K. Aoki. 1981. Cancer deaths by city and county in Japan (1969–1971): A test of significance for geographic clusters of disease. *Social Science and Medicine* 15D:251–258.

Openshaw, S., M. Charlton, C. Wymer, and A. Craft. 1987. A mark I geographical analysis machine for the automated analysis of point data sets. *International Journal of Geographical Information Systems* 1:335–358.

Pisani, J.F., J.J. Angulo, and C.K. Takiguti. 1984. An objective reconstruction of the chain of contagion. *Social Science and Medicine* 18(9):775–782.

Pyle, G.F. 1973. Measles as an urban health problem: The Akron example. *Economic Geography* 49:344–356.

Pyle, G.F. and B.F. Lauer. 1975. Comparing spatial configurations: Hospital service areas and disease rates. *Economic Geography* 51:50–68.

Raine, J.W. 1978. Summarizing point patterns with the standard deviational ellipse. *Area* 10:328–333.

Ribeiro, J.M., C.F. Seulu, T. Abose, G. Kidane, and A. Teklehaimanot. 1996. Temporal and spatial distribution of anopheline mosquitos in an Ethiopian village: Implications for malaria control strategies. *Bulletin of the World Health Organization* 74(3):299–305.

Ricketts, T.C., L.A. Savitz, W.M. Gesler, and D.N. Osborne (Eds.). 1994. *Geographic Methods for Health Services Research: A Focus on the Rural-Urban Continuum.* Lanham, MD: University Press of America.

Ripley, B.D. 1981. *Spatial Statistics.* New York: Wiley.

Rogers, E.M. 1979. Network analysis of the diffusion of innovations. In *Perspectives on Social Network Research,* P.W. Holland and S. Leinhardt (Eds.), pp. 137–167. New York: Academic Press.

Rushton, G. 1975. *Planning Primary Health Services for Rural Iowa: An Interim Report.* Technical Report No. 39. Iowa City: Center for Locational Analysis, Institute of Urban and Regional Research, University of Iowa.

Rushton, G. and P. Lolonis. 1996. Exploratory spatial analysis of birth defect rates in an urban population. *Statistics in Medicine* 15(7/9):717–726.

Rushton, G., R. Krishnamurthy, D. Krishnamurti, P. Lolonis, and H. Song. 1996. The spatial relationship between infant mortality and birth defect rates in a U.S. city. *Statistics in Medicine* 15(17/18):1907–1919.

Schneider, D., M.R. Greenberg, M.H. Donaldson, and D. Choi. 1993. Cancer clusters: The importance of monitoring multiple geographic scales. *Social Science and Medicine* 37(6):753–759.

Scott, A.J. 1970. Location-allocation systems: A review. *Geographical Analysis.* 2:95–119.

Shannon, G.W. and M.P. Cutchin. 1994. General practitioner distribution and population dynamics: Munich, 1950–1990. *Social Science and Medicine* 39 (1):23–38.

Shannon, G.W., R.L. Bashshur, and C.W. Spurlock. 1978. The search for medical care: An exploration of urban black behavior. *International Journal of Health Services* 8(3):519–530.

Smith, A.H., J.D. Kark, J.C. Cassel, and G.F. Spears. 1977. Analysis of prospective epidemiologic studies by minimum distance case-control matching. *American Journal of Epidemiology* 105(6):567–574.

Smith, C.J. 1983. Locating alcoholism treatment facilities. *Economic Geography* 59:368–385.

Smith, C. and R.Q. Hanham. 1981. Any place but here! Mental health facilities as noxious neighbors. *Professional Geographer* 33(3):326–334.

Stimson, R.J. 1983. Research design and methodological problems in the geography of health. In *Geographical Aspects of Health: Essays in Honour of Andrew Learmonth*, N.D. McGlashan and J.R. Blunden (Eds.), pp. 321–334. London: Academic Press.

Tanaka, T., S. Rhu, M. Nishigaki, and M. Hashimoto. 1981. Methodological approaches on medical care planning from the viewpoint of a geographical allocation model: A case study on South Tama District. *Social Science and Medicine* 15D:83–91.

Thomas, R.W. 1979. *An Introduction to Quadrat Analysis.* CATMOG 12. Norwich: Geo Abstracts.

Tinkler, K.J. 1977. *An Introduction to Graph Theoretical Methods in Geography.* CATMOG 14. Norwich: Geo Abstracts.

Unwin, D. 1975. *An Introduction to Trend Surface Analysis.* CATMOG 5. Norwich: Geo Abstracts.

Unwin, D. 1981. *Introductory Spatial Analysis.* London: Methuen.

Waller, L.A. and B.W. Turnbull. 1993. The effects of scale on tests for disease clustering. *Statistics in Medicine* 12(19/20):1869–1884.

Waller, L.A. and G.M. Jacquez. 1995. Disease models implicit in statistical tests of disease clustering. *Epidemiology* 6(6):584–590.

Walsh, S.J, P.H. Page, and W.M. Gesler. 1997. Normative models and healthcare planning: Network-based simulations within a geographic information system environment. *Health Services Research* 32(2):243–260.

Walter, S.D. 1993. Assessing spatial patterns in disease rates. *Statistics in Medicine* 12(19/20):1885–1894.

Wartenberg, D. and M. Greenberg. 1993. Solving the cluster puzzle: Clues to follow and pitfalls to avoid. *Statistics in Medicine* 12(19/20):1763–1770.

White, R.R. 1972. Probability maps of leukemia mortalities in England and Wales. In *Medical Geography: Techniques and Field Studies*, N.D. McGlashan (Ed.), pp. 173–185. London: Methuen.

Williams, G. 1984. Time-space clustering of disease. In *Statistical Methods for Cancer Studies*, R.G. Cornell (Ed.), pp. 167–227. New York: M. Dekker.

Wilson, J.L. 1993. Mapping the geographical diffusion of a Finnish smallpox epidemic from historical population records. *Professional Geographer* 45(13):276–286.

Wojdyla, D., L. Poletto, C. Cuesta, C. Badler, and M.E. Passamonti. 1996. Cluster analysis with constraints: Its use with breast cancer mortality rates in Argentina. *Statistics in Medicine* 15(7/9):741–746.

Chapter Three

Geographic Information Systems: Medical Geography

PRESCRIPT

This chapter originally appeared as an article entitled "Geographic Information Systems and Health: An Educational Resource" in the March/April 1995 issue of the *Journal of Geography* (Albert, Gesler, and Wittie, 1995). It is reprinted here with the permission of the National Council for Geographic Education. The article chronicles the early years (through 1993) of the diffusion of geographic information systems into medical geography and related disciplines. It documents a small but vibrant body of research that was grappling with the introduction of GIS into the realms of health and disease. While some scholars were optimistically urging usage of this emerging technology, others were advocating caution before jumping on the GIS bandwagon. All the while, a handful of investigators began to develop and operationalize applications of geographic information systems with a specific focus on health and/or disease.

INTRODUCTION

Past articles in the *Journal of Geography* have encouraged the teaching of geography through the use of computer technology (Fitzpatrick, 1993), including Geographic Information System (GIS) technology; promoted GIS as an educational tool that stimulates creative thought and problem solving (White and Simms, 1993); proposed a balanced curriculum that incorporates GIS, remote sensing, and cartography with human, physical, and regional courses (King, 1991); and delineated software and hardware configurations for cartographic and spatial analysis laboratories in an educational environment (Walsh, 1992). This article provides a review of the literature on the applications of GIS to studies of health care and disease ecology in the social sciences. Although there have been numerous textbooks and research articles on GIS (Young, 1986; Aronoff, 1989; Cowen, 1990; Marble, 1990; Peuquet and Marble, 1990; Star and Estes, 1990; Antenucci et al., 1991; ICMA, 1991; Martin, 1991; Parr, 1991; Korte, 1992), references in the literature to the use of GIS to study health risks or "access-to-care" problems are sparse. Whereas there has been a strong emphasis on the use of GIS in physical geography, human geographers have largely overlooked its potential. This imbalance is gradually changing as software and hardware prices drop and more students are trained in data management, computer technology, and cartographic principles. Consequently, a review is in order to illustrate the changes which are occurring in the use of GIS particularly as it relates to spatial aspects of health.

This article is intended as a bibliographic resource for university geography teachers and students. It is aimed primarily at those who teach courses in medical geography and GIS, but could be useful for a variety of physical and human geography courses at both the undergraduate and graduate levels. In the first section, we will very briefly discuss the various methods of defining and naming a GIS. In the second section, we review the literature on using GIS in medical studies. This section follows a progression from studies that advocate the use of GIS, to cautionary literature, to preliminary investigations, and finally to actual applications. The focus is on the call for "the application of geographical concepts and techniques to health-related problems" (Hunter 1974, p. 3) within a GIS environment.

DEFINITION OF GIS

Geographical information systems (GISs) have been defined in different ways, based on their functions, basic components, and uses. For instance, Antenucci et al. (1991, p. 281) defined GIS as a "computer system that stores and links nongraphic attributes or geographically referenced data with graphic map features to allow a wide range of information processing and display operations, as well as map production, analysis, and modeling." Parr (1991, p. 2) defined a GIS according to its basic components, which include: (1) data input and editing, (2) data management, (3) data query and retrieval, (4) analysis, modeling, and synthesis, and (5) data display and output functions. Cowen (1990) viewed GIS as the integration of spatial data for decision-support systems.

A number of terms are used which indicate the type of user employing a GIS, such as land information system (LIS), land records information system (LRIS), urban information system (URIS), environmental resource information system (ERIS), cadastral (legal registration of land parcels) information system (CAIS), geographic information processing (GIP), and Geomatics (*Geomatique*) in Canada (Taylor, 1991). In the health sciences, some specific references to a GIS are worth noting. Openshaw et al. (1987; 1988) developed a geographical analysis machine (GAM) that combines statistics and a GIS to determine the significance of leukemia clusters. Twigg (1990) referred to a health information system (HIS), and Wrigley (1991) discussed a health agency geographic information system (HAGIS).

THE LITERATURE ON USING GIS IN MEDICAL STUDIES

Background

A very diverse collection of professional and academic journals contain a relatively small number of articles on using a GIS to study either health care or disease. The effort required to collect even a limited number of references for such a review is enormous. Bibliographic searches using computer systems and the manual scanning of abstracts provided an excellent beginning; however, these search methods alone do not give a complete picture. The task then became one of sifting through the "gray literature," such as the trade journals, association yearbooks, and conference and workshop proceedings to find more obscure references. Even this more comprehensive bibliographic search was deficient in several respects. First, a search for project documents from local agencies, commissions, and authorities was futile. Second, foreign professional and academic journals, trade and association journals, and conference proceedings often have a limited circulation within the United States. Third, interest in GIS for health or diseases research stretches across numerous disciplines, including geography, urban and regional developmental studies, planning, geology, epidemiology, environmental health, health services research, social medicine, and oncology.

Geographers have been in the forefront of using GIS in medical research. For example, at the 1993 Annual Meeting of the Association of American Geographers in Atlanta, for example, eight papers were presented on the use of a GIS in health research. These papers discussed emergency response (Lewis, 1993), hazardous waste assessment (Fowler et al., 1993; Padgett, 1993); pesticide exposure (McDonald, 1993; Tiefenbacher, 1993); powerline corridors and negative health impacts (Whately, 1993); female breast and non-Hodgkins lymphoma cancers (Wittie et al., 1993); and radon exposure (Fandrich, 1993). Some of these papers will evolve into publishable research articles that will encourage other medical geographers to use GIS technology.

Categorizing the Literature

This literature review is organized into four basic groups (Figure 3.1). Articles in the first group see the potential of GIS as "an opportunity we should seize" (Fishenden 1991, p. 127). Articles in the second group argue for using caution before seriously contemplating the use of a GIS because disease and health data are often inaccurate and therefore incompatible with the precision of a database management system. Articles discussed in the third group suggest some untested preliminary uses of a GIS in health and disease research; such research sets the stage for later research projects. Articles in the fourth group provide several applications of GISs in addressing the spatial aspects of health and disease research. Each group is important; academic and professional exchange through journals, conferences, and workshops provide a dialogue between the four groups. Our purpose is to facilitate the dialogue among the groups in the form of an accessible literature review for the social sciences.

THE POTENTIAL FOR USING GIS

The potential of GIS in medical research has been espoused by many scholars. Verhasselt (1993) discussed applications in medical geography. This new tool, she

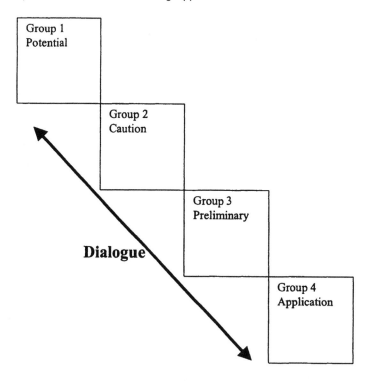

Figure 3.1. Four basic literature groups in geomedical applications of GIS.

says, facilitates hypothesis-generating because of its capability to: (1) overlay and integrate spatial information, and (2) substantiate quantitative analyses in disease ecology and health care delivery through its capability to handle large amounts of data. While this technology requires strict measures to ensure data quality, it also "opens up new possibilities in ecological associative analysis" (Verhasselt, 1993, p. 121). Scholten and de Lepper (1991) supported the idea of using a GIS approach because of its potential value to improve public health and understand environmental risk. While these authors cite the ability to manage, manipulate, and analyze large quantities of spatial data more quickly and with less effort than conventional methods, they also recognize that the need for expertise limits the number of studies employing a GIS. However, a multidisciplinary approach could alleviate this problem somewhat. Twigg (1990) also recognized the value of a GIS in the management and analysis of health care data. Her main concern resided with the necessity for timely, spatially-referenced, accurate data. In the case of secondary data, which is often the only available health and disease data, this can be problematic.

    Much of the research on the potential of GIS in medical geography stems from discussions regarding the National Health Service (NHS) of the United Kingdom (Curtis and Taket, 1989; Nichols, 1991; Gould, 1992). Wrigley (1991, p. 7) saw GIS as "critical to the operationalization of the internal market" (e.g., hospitals form internal markets by competing for service contracts from among groups of general practitioners or District Health Authorities within the NHS). Wrigley stated that a health agency geographic information system (HAGIS) could be developed to: (1) manage resources, (2) monitor family practitioners, and (3) assist epidemiological studies. Gould (1992) saw enormous potential for the use of GIS and geographically based analysis in the NHS. Although some District Health Authorities possess GIS capabilities, ineffective use of GISs reduced most analyses to computer-assisted cartography. Fishenden (1991) promoted the use of statistics and GIS in the Health Education Authority, stating that GIS technology must be incorporated to meet the challenge of improving the health of the population of the United Kingdom. Others also envisioned GIS within the future of the NHS (Mohan and Maguire, 1985). The greatest potential for GIS lies in its ability to cross geopolitical boundaries or facilitate small area investigations. Wider usage among NHS planners depends on the development of more user-friendly and inexpensive software/hardware combinations since both finances and expertise remain scarce.

RESERVATIONS ABOUT GIS USE

This segment of the literature argues for caution in using GISs. One general concern lies in the lack of a "universally accepted definition" (Taylor 1991, p. 5). The second concern relates to the issue of the suitability of a GIS to answer certain research questions. Furthermore, research tools and techniques should not overshadow the research question. In some instances, using a GIS evokes more questions than it does answers, which of course supports hypothesis generation, but also possibly fails to answer the original question (Heywood, 1990). Thus, a GIS necessitates an intelligent approach to ensure sound analysis. Matthews (1990) concurred with this thought as he argued for a critical approach to using a GIS in epidemiology. A third problem resides in the data-intensive nature of a GIS. The requisite disease and health information may be inaccessible. The underlying reasons for inaccessible data include not only its expense but also the political nature of establishing data sets (Taylor and Overton, 1991). Often data are difficult to access because those in control guard their release. If they are released, they may not be in a usable form. Sometimes secondary data are placed within a GIS environment without acknowledging the limitations of the data set. The importance of recognizing data limitations prior to analysis cannot be overemphasized. If the use of GIS is to reach its full potential, then the academic community must understand its basic concepts and this is not likely to occur for a whole host of reasons, especially the issue of inaccurate or misused data (Heywood, 1990).

Twigg (1990) noted some problems of using routinely collected data in GIS health care research. She finds incomplete and inaccurate data sets that have been compiled for large administrative units. In addition, data sets typically come from different organizations in a wide variety of incompatible spatial units (Heywood, 1990). Similarly, Matthews (1990) found numerous problems concerning data collection and verification. He recommended multivariate statistics over GIS in issues of disease causation because of scale issues and small numbers in epidemiological research.

PRELIMINARY INVESTIGATIONS

This literature suggests the use of a GIS for medical research, but does not actually carry out empirical studies. Lam (1986) analyzed the geographical patterns of stomach and esophageal cancer for males and females for the period of 1973–1975 for the provinces, municipalities, and autonomous regions of China. High rates of stomach cancer occurred in the west and northwest provinces of Xizang, Qinghai, and Ningxia, whereas high rates of esophageal cancer surfaced in the north-central and eastern areas of Jiangsu, Shanxi, and Henan. Lam (1986) believed that a GIS could facilitate a more efficient investigation of the environmental influences on the risk of cancer, given the enormous size of a database needed to facilitate such research.

Wartenburg (1992) proposed using a GIS for: (1) lead exposure prediction; (2) mapping cases of lead exposure; and (3) validation of exposure prediction with incidence data. He illustrated these steps by using a series of four-by-four matrices (cells) as a base map for a hypothetical community, each cell serving as a neighborhood. The socio-economic status of each neighborhood (e.g., 0–3 with 0 = high and 3 = low) was shown in the cells of the first map. The second map recorded soil contamination scores (e.g., 0 = clean and 3 = contaminated) and air contamination scores (0 = no effect and 3 (equals) maximum effect) for each of the cells. The cell scores for socio-economic, soil and air contamination were added to produce a lead-exposure risk score for each cell; the results produced a third map. Actual lead-exposure cases can be shown with point symbols on another four-by-four matrix. The number of observed lead-exposure cases per cell (neighborhood) can be compared to the predicted lead-exposure risk score for validation. Neighborhoods showing low correspondence between observed and predicted cases can be targeted for further investigation or the prediction model can be refined with additional risk-factor variables.

Stallones et al. (1992) suggested the use of a GIS to assess reproductive outcomes in an area surrounding hazardous waste sites. Three thematic overlays—(1) a base map showing the landfill and all residences with water wells, (2) the zone of aquifer contamination, and (3) water wells drawing water from contaminated wells—were used to create a composite map of the area for comparison to reproductive outcomes measures (e.g., low birth weight). In this manner, a surveillance program can monitor reproductive outcomes and environmental risk to formulate health assessments of neighborhoods in close proximity to hazardous waste sites. In the same vein, Estes et al. (1987) discussed monitoring and managing hazardous waste sites using GISs.

APPLYING GIS IN DISEASE AND HEALTH RESEARCH

The literature discussed in this section shows how GIS technology has been employed to answer specific research questions and problems in health care delivery and disease ecology (Table 3.1).

Table 3.1. Applications of Geographic Information Systems to Health Research.

Authors	Application	Geography	Variables	Concept Illustrated	Software
Tyler, 1990	Search and rescue	Mirco scale San Francisco and Alameda County	Vehicle registration Power lines	Database query Reports	Unspecified
Van Creveld, 1991	Automated emergency response	West Midlands 345 square miles	Incoming reports Real-time tracking of 24 rescue vehicles	Network routing	West Midland's Command and Control Program
Fost, 1990	AIDS surveillance	San Francisco	Street network Questionnaire data	Database queries	MapInfo
Twigg, 1990	Assessment of GPs and clinics relative to school catchment districts	Enumeration district approximation using Thiessen polygons	Households without WC and bath GP surgeries and clinics School catchment districts	Thiessen polygons	ARC/INFO
Zwarenstein et al., 1991	Measuring accessibility	Defining hospital catchment areas using Thiessen polygons	Persons/bed ratios	Thiessen polygons	ARC/INFO
Moore, 1991	Degree of public exposure Assess health risks from emissions	Isopleths of cancer risk	Carcinogenic and noncarcinogenic health risks (facility specific)	Isopleths Modeling	PC ARC/INFO

Table 3.1. Applications of Geographic Information Systems to Health Research. (Continued)

Authors	Application	Geography	Variables	Concept Illustrated	Software
Guthe et al., 1992	Prediction of high blood lead levels among children	Identification of high-risk census tracts	Blood screening records Local sources of industrial and hazardous wastes Traffic volume	Overlay operations Modeling	ARC/INFO
Fitzpatrick-Lins et al., 1990	Regional variations in radon potential	Within county	Underlying geology Soils Indoor radon data	Overlay operation Exploratory data analysis	ARC/INFO
Openshaw et al., 1990	Significance of cancer clusters	Point locations	Childhood leukemia incidence	Neighborhood operation—search using radii	GAM
Solarsh and Dummann, 1992	Measles surveillance	Subdistrict (Edendal Health Ward, southern Natal)	Hospital inpatient data	Scale	DBase III PLUS Epi Info (v. 3 CDC) Harvard Graphics

Emergency Response

On 17 October 1989, the California Bay Area was hit with a serious earthquake registering 7.1 on the Richter scale. Alameda County had a GIS in operation and dedicated it to their emergency response activities that included the identification of damaged vehicles and downed power lines. The GIS facilitated the management of incoming reports from rescue crews and provided a rapid and concerted emergency response. At the same time, the California Highway Patrol set up a makeshift command center beside the rubble of a collapsed section of the Nimitz Freeway. Using Alameda's GIS software and a portable computer, officers were able to systematically monitor the search-and-rescue operation. Reports were issued periodically on each 80-foot section to show the location of trapped vehicles (Tyler, 1990).

In another example, West Midlands Ambulance Service in the United Kingdom is moving toward a comprehensive GIS for emergency response. In 1991, West Midland's computerized Command-and-Control program could locate the nearest 24 rescue vehicles, determine their current status, and display this information on a monitor. A tracking system will be added, using radio beacons, to determine the optimum vehicle for dispatch to an accident. The assigned rescue vehicle appears on a monitor as a moving icon. The system will be able to: (1) output immediate reports regarding the routing of a vehicle, and (2) tabulate rescue responses by type of incident in any given time period (Van Creveld, 1991; see also Dunn and Newton, 1992, for optimal routing algorithms for emergency planning).

AIDS Prevention

San Francisco's Youth Environment Study (YES), a local nonprofit organization, used MapInfo (GIS) in its fight against AIDS. YES field workers canvassed the streets searching for intravenous drug users. Once a user was found, field workers distributed vials of bleach for needle sterilization and condoms as protection against the virus, and administered a 45-page questionnaire to intravenous drug users. The data from the questionnaire were brought into MapInfo's software. Through a standard query language (SQL), the software permitted YES to access the data base for specific information on spatial (location) and nonspatial (age, ethnic group, etc.) attributes of intravenous drug users who use and share unclean needles. The age and ethnic group of intravenous drug users was matched to those of field workers in order to reduce communication barriers. In this manner, YES was able to concentrate resources on the most susceptible of this high-risk population (Fost, 1990).

Catchment Area Studies

ARC/INFO, which is a GIS developed by Environmental Systems Research Institute (ESRI) has been used in at least two projects involving catchment areas. In the following examples, the term catchment refers to an area from which a school, hospital, or other service institution draw most or all of their students, patients, or clientele respectively. Twigg (1990) examined the following variables in school catchment areas in Portsmouth, England: (1) the number of households without bathroom facilities; (2) general practitioner (GP) surgeries; and (3) health clinics. Twigg overcame several problems of existing data sources by aligning the schools' perimeter arcs with more accurate boundary files containing road and railroad networks. Thiessen polygons were used to approximate the boundaries of enumeration districts, which ranged in size from 150 to 500 persons. This GIS procedure saved time and money by eliminating the need for digitizing these small spatial units. The approximated enumeration districts were then used as a base to (1) map socio-economic variables such as the number of households without bathroom facilities; (2) generate catchment areas around general practitioner surgeries; (3) identify schools located within a 300 meter buffer of a general practitioner surgery; and (4) show school and clinic locations relative to catchment areas. Twigg found the use of GIS for health research useful despite the problem of acquiring accurate data at appropriate spatial scales.

Zwarenstein et al. (1991) found ARC/INFO GIS useful in analyzing the affect of removing race restrictions on hospitals in Natal/KwaZulu, South Africa, in 1985. Again, Thiessen polygons via ARC/INFO software were used to represent catchment areas. Three maps were produced using Thiessen polygons to define catchment areas. These included: (1) white referral and general hospitals, (2) black referral and general hospitals, and (3) all referral and general hospitals (race restriction removed). The results indicated that even with the removal of race restrictions on hospitals the population/bed ratio did not significantly improve for blacks.

Monitoring and Surveillance Including Aspects of Modeling and Simulations

In 1987 California passed an air toxics "hot spots" act which called for the identification of carcinogenic and noncarcinogenic health risks of facility-specific air toxics emissions. Two initial objectives of the programs were to determine the degree of public exposure and to assess the potential health risk from air toxics emissions from a facility. These objectives were accomplished using PC ARC/INFO. In one part of the research, isopleths were drawn around a facility to indicate the worst possible excess cancer risk due to operational emissions from a facility (Moore, 1991).

Guthe et al. (1992) conducted a pilot project to compare the expected versus actual spatial pattern of high blood lead among children of Newark, East Orange, and Irving, New Jersey, using a GIS. The following data bases were brought together for the purpose of predicting spatial patterns of lead exposure from known risk factors: (1) U.S. Census TIGER Line files; (2) blood screening records from the New Jersey Department of Health; (3) local sources of industrial and hazardous waste from the New Jersey Department of Environment Protection and Energy; and (4) traffic counts from the New Jersey Department of Transportation. Noticeable differences existed between the observed and expected spatial patterns of lead exposure. These differences suggested that additional variables should be incorporated into the model for more accurate lead-exposure prediction.

Solarsh and Dammann (1992) brought together dBase III PLUS, Epi Info, (an epidemiological data analysis program from the Centers for Disease Control) and Harvard Graphics to produce a community pediatric information system (CPIS) to monitor longitudinal child health trends (e.g., measles surveillance) in the Edendal Health Ward in southern Natal, South Africa. Given hospital inpatient data (e.g., date of admission, sex, age, vaccination status, etc.), the community pediatric information system can "pinpoint" areas experiencing a rapid increase in measles incidence. This customized system offered public health officials the appropriate spatial information to focus efforts during periods of measles outbreaks.

Cancer-Related Research

Fitzpatrick-Lins et al. (1990), using exploratory data analysis and GIS, found that radon potential was high for the piedmont upland of Fairfax County, Virginia. They discovered the metamorphic rock of the piedmont was associated with a high radon potential, while soils on Triassic sandstone and shale had a mediating influence on potential radon levels. Radon overexposure is responsible for approximately 20,000 lung cancer deaths in the United States each year (Fitzpatrick-Lins et al., 1990).

Openshaw et al. (1987; 1988) developed a geographical analysis machine (GAM) that was used to test the significance of childhood leukemia cancer clusters. Within the GIS component of GAM, a grid of points was superimposed over the study area. Each point of the grid was used as a center for a set of concentric circles. The age-sex adjusted incidence rate of childhood leukemia was calculated for each circle and tested for significance based on Poisson probabilities. The advantages of this procedure were the use of disaggregate cancer incidence data and the fact that potential cancer clusters were tested from multiple adjacent locations on the grid (Thomas, 1992).

CLASSROOM LINKAGE

This literature review can be used as supplementary reading to geographic information systems, medical geography, and health policy and administration courses. Textbooks used in these courses often fail to mention or inadequately address possible applications of GIS with medical geography and health services research. For example, Aronoff (1989) mentions some examples of GIS applications in his introductory GIS textbook. He cites applications for agriculture and land use planning, forestry and wildlife management, archaeology, geology, and municipal applications. However, there are no examples that highlight applications of GIS to spatial aspects of health and disease.

This review of the literature can be useful in a GIS course in several ways. First, this review can familiarize GIS instructors, most of whom do not have a strong medical geography orientation, with potential applications for investigating spatial aspects of health and disease. This is especially important if the instructor has a large number of students with interests in medical or human geography. Instructors unfamiliar with specific applications of GIS to issues of health and disease often, understandably, fall short when creating medical geography scenarios to illustrate GIS concepts. Second, the instructor can use some of the research articles cited in this review to illustrate certain concepts and functions of geographic information systems. For example the following GIS capabilities can be referenced with specific research (Table 3.1): database attribute queries and report generation (Tyler, 1990), database queries (Fost, 1990), overlay operations (Fitzpatrick-Lins et al., 1990; Guthe et al., 1992), exploratory data analysis (Fitzpatrick-Lins et al., 1990), Thiessen polygons (Twigg, 1990; Zwarenstein et al., 1991), isopleths (Moore, 1991), radii searches and cluster significance (Openshaw et al., 1987), network routing (Van Creveld, 1991), scale (Solarsh and Dammann, 1992), and modeling (Moore, 1991; Guthe et al., 1992). Third, most GIS courses have a term project of some sort; this review provides students with ideas and direction that can be used to develop a project more in line with improving on previous research or identifying new avenues of research.

This paper would also be useful in a medical geography course to illustrate geographic tools, techniques, and technology. Geographic information systems are standard and accepted in this era and most subdisciplines in geography have found numerous applications for GIS technologies. Thus most geographers will find it a benefit to have at least a rudimentary knowledge of what constitutes a GIS (e.g., definition and components) and of previous applications in their specialty.

Finally, this review has a potential use among students and instructors of courses outside the discipline of geography as well. As more disciplines, especially those such as epidemiology, health policy administration, and environmental science entertain the use of geographic information systems, there is need for a concise literature review to familiarize students and instructors of these disciplines with some of the leading scholars, articles, textbooks, journals, and software supporting the integration of GIS with research on health and disease. Geographic information systems can provide a connection that links geographers with other specialists across the disciplines. This linkage of geographers with other specialists is an advantageous situation for all concerned and one that will provide a balancing and sensitivity for disciplinarian perspectives and strengthen research overall.

CONCLUSION

The literature on GIS in health and disease research is still somewhat immature, but is developing more rigor in a balanced fashion. Ninety percent of the references cited in this paper were published in 1990 or later; based on the evolution of GIS for medical studies there could be an explosion of research in the mid and late 1990s. Quantitatively, the number of research articles and projects is very small, but qualitatively the literature as a whole is sound. The literature is balanced in that four distinct components represent potential, cautionary, preliminary, and practical approaches to GIS. Each component is invaluable in furthering the development of GIS in health and disease research. The literature presents the student and instructor enough of a glimpse of the advantages and disadvantages of GIS to develop a realistic appraisal of the possibilities of using GIS in medical studies. Enough examples have been described here to provide teachers with material for a classroom unit on using (1) GIS in a medical geography class; (2) medical applications in a GIS class; or (3) using GIS in classes which have a health and disease component.

POSTSCRIPT

Since the publication of "Geographic Information Systems and Health: An Educational Resource," there have been dynamic changes with respect to software usage and the components of research. First, desktop mapping and GIS software are becoming more user friendly. Take for example the current popularity of such products as MapInfo™ and ArcView GIS, and others. Second, the explosion of GIS research highlighting geomedical applications predicted for the late 1990s is upon us. Electronic searches of MedLine and other citation/abstract databases are yielding more and more "hits." Third, in terms of the four basic literature components, groups 1 (Potential) and 3 (Preliminary) have lost their raison d'être. The potential group is no longer relevant as it is now generally accepted that GIS offers tremendous benefits in terms of managing and analyzing spatial databases. No one needs to be convinced any longer! The preliminary group (research setting up mock projects) is also phasing out as GIS software and hardware have becoming less expensive and technical expertise more available. Group 3, or research advocating caution, has tempered its concerns. Initial objections about data quality, scale issues, fears that technology might drive research, and other problems inherent in analyzing patterns of diseases (latency, mobility, multiple causation) have been well said and noted by serious practitioners and academics. The dominant research component is within group 4, or applications. The number and type of applications have increased dramatically since this article was originally published. This will become evident as we explore a plethora of GIS applications in Chapter 4 (health services research), Chapter 5 (environmental and public health), and Chapter 6 (infectious diseases).

REFERENCES

Albert, D.P, W.M. Gesler, and P.S. Wittie. 1995. Geographic information systems and health: An educational resource. *Journal of Geography* 94(2):350–356.

Antenucci, J.C., K. Brown, P.L. Croswell, M.J. Kevany, and H. Archer. 1991. *Geographic Information Systems: A Guide to the Technology*. New York: Van Nostrand Reinhold.

Aronoff, S. 1989. *Geographic Information Systems: A Management Perspective*. Ottawa: WDL Publications.

Curtis, S.E. and A.R. Taket. 1989. The development of geographical information systems for locality planning in health care. *Area* 21(4):391–399.

Cowen, D.J. 1990. GIS versus CAD versus DBMS: What are the differences? In *Introductory Readings in Geographic Information Systems*, D.J. Peuquet and D.F. Marble (Eds.), pp. 52–61. London: Taylor & Francis.

Dunn, C.E. and D. Newton. 1992. Optimal routes in GIS and emergency planning applications. *Area* 24(3):259–267.

Estes, J.E., K.C. McGwire, G.A. Fletcher, and T.W. Foresman. 1987. Coordinating hazardous waste management activities using geographical information systems. *International Journal of Geographical Information Systems* 1(4):359–377.

Fandrich, J.E. 1993. Using a GIS to evaluate radon potential and its effect on housing. *Abstracts: 89th Annual Meeting of the Association of American Geographers.* Washington, DC: Association of American Geographers.

Fishenden, J. 1991. Towards more healthy living. In *Geographic Information 1991: The Yearbook of the Association for Geographic Information,* pp. 126–127. London: Taylor & Francis.

Fitzpatrick, C. 1993. Teaching geography with computers. *Journal of Geography* 92(4):156–159.

Fitzpatrick–Lins, K., T.L. Johnson, and J.K. Otton. 1990. Radon potential defined by exploratory data analysis and geographic information systems. *U.S. Geological Survey Bulletin* 1908:E1–E10.

Fost, D. 1990. Using maps to tackle AIDS. *American Demographics* 12(4):22.

Fowler, G.L., M.A. Lazar, R.K. Dillow, J.W. Bash, and R.J. Dutton. 1993. A hazardous substances GIS for public health assessment. *Abstracts: 89th Annual Meeting of the Association of American Geographers.* Washington, DC: Association of American Geographers.

Gould, M.I. 1992. The use of GIS and CAC by health authorities: Results from a postal questionnaire. *Area* 24(4):391–401.

Guthe, W.G., R.K. Tucker, E.A. Murphy, R. England, E. Stevenson, and J.C. Luckhardt. 1992. Reassessment of lead exposure in New Jersey using GIS technology. *Environmental Research* 59(2):318–325.

Heywood, I. 1990. Geographic information systems in the social sciences. *Environment and Planning A* 22(7):849–854.

Hunter, J.M. 1974. The challenge of medical geography. In *The Geography of Health and Disease: Papers of the First Carolina Geographical Symposium.* J.M. Hunter (Ed.). Chapel Hill: Department of Geography, University of North Carolina.

King, G.Q. 1991. Geography and GIS technology. *Journal of Geography* 90(2):66–72.

Korte, G.B. 1992. *The GIS Book: A Practitioner's Handbook,* 2nd ed. Sante Fe, NM: OnWord Press.

Lam, N.S. 1986. Geographical patterns of cancer mortality in China. *Social Science and Medicine* 23(3):241–247.

Lam, N.S. and D.A. Quattrochi. 1992. On the issues of scale, resolution, and fractal analysis in the mapping sciences. *Professional Geographer* 44(1):88–98.

Lewis, L. 1993. Using GIS for development of a rural 9-1-1 system. *Abstracts: 89th Annual Meeting of the Association of American Geographers.* Washington, DC: Association of American Geographers.

Marble, D.F. 1990. Geographic information systems: An overview. In *Introductory Readings in Geographic Information Systems,* D.J. Peuquet and D.F. Marble (Eds.), pp. 8–17. London: Taylor & Francis.

Martin, D. 1991. *Geographic Information Systems and Their Socioeconomic Applications.* London: Routledge.

Matthews, S.A. 1990. Epidemiology using a GIS: The need for caution. *Computers, Environment and Urban Systems* 14(3):213–221.

McDonald, R.E. 1993. Delaware watershed pesticide impact assessment model. *Abstracts: 89th Annual Meeting of the Association of American Geographers*. Washington, DC: Association of American Geographers.

Mohan, J. and D. Maguire. 1985. Harnessing a breakthrough to meet the needs of health care. *Health and Social Service Journal* 95(4947):580–581.

Moore, T.J. 1991. Application of GIS technology to air toxics risk assessment: Meeting the demands of the California air toxics "Hot Spots" Act of 1987. *GIS/LIS '91 Proceedings* 2:694–714.

Nicol, J. 1991. Geographical information systems within the National Health Service: The scope for implementation. *Planning Outlook* 34(1):37–42.

Openshaw, S., M. Charlton, C. Wymer, and A. Craft. 1987. A mark I Geographical Analysis Machine for the automated analysis of point data sets. *International Journal of Geographical Information Systems* 1(4):335–358.

Openshaw, S., A.W. Craft, M. Charlton, and J.M. Birch. 1988. Investigation of leukaemia clusters by use of a Geographical Analysis Machine. *Lancet* February 6:272–273.

Padgett, D.A. 1993. Spatial correlation between racial/socioeconomic class and hazardous waste releases. *Abstracts: 89th Annual Meeting of the Association of American Geographers*. Washington, DC: Association of American Geographers.

Parr, D.M. 1991. *Introduction to Geographic Information Systems Workshop*. The Urban & Regional Information Systems Association.

Peuquet, D.J. and D.F. Marble 1990. *Introductory Readings in Geographic Information Systems*. London: Taylor & Francis.

Public Technology, Inc., Urban Consortium for Technology Initiatives, and International City Management Association. 1991. *The Local Government Guide to Geographic Information Systems: Planning and Implementation*. Washington, DC: Public Technology, Inc. and International City Management Association.

Scholten, H.J. and M.J. de Lepper. 1991. The benefits of the application of geographical information systems in public and environmental health. *World Health Statistics Quarterly* 44(3):160–170.

Stallones, L., J.K. Berry, and J.R. Nuckols. 1992. Surveillance around hazardous waste sites: Geographic information systems and reproductive outcomes. *Environmental Research* 59(1):81–92.

Star, J. and J. Estes. 1990. *Geographic Information Systems: An Introduction*. Englewood Cliffs, NJ: Prentice Hall.

Solarsh, G.C. and D.F. Dammann. 1992. A community paediatric information system: A tool for measles surveillance in a fragmented health ward. *South African Medical Journal* 82(2):114–118.

Taylor, D.R.F. (Ed.). 1991. *Geographic Information Systems: The Microcomputer and Modern Cartography*. Oxford, England: Pergamon Press.

Taylor, P.J, and M. Overton. 1991. Further thoughts on geography and GIS. *Environment and Planning A* 23:1087–1094.

Tiefenbacher, J.P. 1993. Application of GIS to the analysis of residential exposure to airborne pesticide drift: Prospects and needs. *Abstracts: 89th Annual Meeting of the Association of American Geographers*. Washington, DC: Association of American Geographers.

Thomas, R. 1992. *Geomedical Systems: Intervention and Control*. London: Routledge.

Twigg, L. 1990. Health based geographical information systems: Their potential examined in the light of existing data sources. *Social Science and Medicine* 30(1):143–155.

Tyler, S. 1990. Computer assistance for the California earthquake rescue effort. *The Police Chief* 57(3):42–43.

Van Creveld, I. 1991. Geographic information systems for ambulance services. In *Geographic Information 1991: The Yearbook of the Association for Geographic Information,* pp. 128–130. London: Taylor & Francis.

Verhasselt, Y. 1993. Geography of health: Some trends and perspectives. *Social Science and Medicine* 36(2):119–123.

Walsh, S.J. 1992. Spatial education and integrated hands-on training: Essential foundations of GIS instruction. *Journal of Geography* 91(2):54–61.

Wartenberg, D. 1992. Screening for lead exposure using a geographic information system. *Environmental Research* 59(2):310–317.

Whately, C.O. 1993. Northcentral Florida overhead powerline corridor study. *Abstracts: 89th Annual Meeting of the Association of American Geographers.* Washington, DC: Association of American Geographers.

White, K.L. and M. Simms. 1993. Geographic information systems as an educational tool. *Journal of Geography* 92(2):80–85.

Wittie, P., W. Gesler, S. McKnight, H. Swofford, and T. Fowler. 1993. Using GIS to create cancer incidence data sets. *Abstracts: 89th Annual Meeting of the Association of American Geographers.* Washington, DC: Association of American Geographers.

Wrigley, N. 1991. Market-based systems of health-care provision, the NHS Bill, and geographical information systems. *Environment and Planning A* 23(1):5–8.

Young, J.A.T. 1986. *A U.K. Geographic Information System for Environmental Monitoring, Resource Planning & Management Capable of Integrating & Using Satellite Remotely Sensed Data.* Nottingham: Remote Sensing Society.

Zwarenstein, M., D. Krige, and B. Wolff. 1991. The use of a geographical information system for hospital catchment area research in Natal/KwaZulu. *South African Medical Journal* 80(10):497–500.

Chapter Four

Geographic Information Systems: Health Services Research

The use of geographic information systems (GIS) within health services research offers exciting potentials. This chapter examines the intersection of GIS and health services research (HSR). Our goals are twofold: (1) to compile and summarize existing literature on GIS *and* HSR, and (2) to assess the degree that HSR utilizes the full potential of GIS. Readers might also want to consult other reviews that focus on health care planning, marketing, and delivery (Barnes and Peck, 1994), geomedical (Albert, 1994; Albert et al., 1995; Albert, 1997a; Croner et al., 1996), public and environmental health (Scholten and de Lepper, 1991; Briggs and Elliott, 1995), and epidemiological (Glass et al., 1993; Mott et al., 1995; Pan American Health Organization, 1996) applications of geographic information systems.

This chapter begins by presenting definitions for geographic information systems and health services research. Some basic concerns about using GIS are briefly noted before describing specific applications. Next, the GIS applications have been organized using the following major divisions: physician distributions; hospitals and other health care facilities; and monitoring, surveillance, and emergency planning. Finally, a discussion follows that critiques the HSR/GIS applications in terms of a four-group classification of GIS functions. This is done to assess the extent to which health service researchers have utilized the potential of GIS.

DEFINITIONS

Geographic Information Systems

GIS is an acronym for geographic information systems. The term GIS is near generic in usage, however, variant acronyms describe similar information systems. For example, health information systems (HIS) and health agency information system (HAIS) (Twigg, 1990; Nicol, 1991; Wrigley, 1991) describe geographic information systems processing health care utilization and resource databases, respectively. There is no universal definition of GIS (Taylor 1991, p. 5); however, experts espouse numerous renditions, including:

- manual or computer databases that store and manipulate geographic data (Aronoff, 1989);
- "computer system of hardware and software that integrates graphics with databases and allows for display, analysis, and modeling" (Public Technology et al., 1991);
- "computer system that stores and links nongraphic attributes or geographically referenced data with graphic map features to allow a wide range of information processing and display operations, as well as map production, analysis, and modeling" (Antenucci et al., 1991, p. 281);
- integration (via a polygon overlay process) of spatial data for decision-support systems (Cowen, 1990);
- data input and editing; data management; data query and retrieval; analysis, modeling, and synthesis; data display and output functions (Parr 1991, p. 2);
- the "four Ms"—measurement, mapping, monitoring, and modeling (Star and Estes, 1990).

These excerpts define geographic information in terms of databases, integration of spatial data, software and hardware, process (i.e., input, editing, management, query and retrieval, etc.), and functional capabilities (i.e., the "four Ms"). Regardless of these definitional nuances, geographic information systems must support users in "solving complex planning and management problems" (Antenucci et al., 1991).

GIS analytic capabilities have been classified into four groups of functions (Aronoff, 1989). These include: (1) maintenance and analysis of the spatial data, (2) maintenance and analysis of the attribute data, (3) integrated analysis of the spatial and attribute data, and (4) cartographic output formatting. Group 1, maintenance and analysis of the spatial data, includes such functions as format transformations, geometric transformations, transformation between map projections, conflation, edge matching, editing of graphic elements, and line coordinate thinning. Group 2, maintenance and analysis of the attribute data, includes attribute editing and attribute query functions. Group 1 and 2 functions, although important, address technical issues of concern to software technicians and system analysts. More pertinent from an investigator's frame of reference is the potential to integrate the analysis of *spatial* and *attribute* data. Herein, group 3 functions provide for the integration of spatial and attribute data through retrieval/classification/measurement, overlay operations, neighborhood operations, and connectivity functions. Group 4, cartographic output formatting, includes such functions as map annotation, text labels, texture and line styles, and graphic symbols for customizing figures, charts, and maps. In a later section, this four-group classification of GIS functions provides an organizing framework to assess the degree that health services research has realized the potential of GIS.

Finally, note is given to one debate raging within professional and academic circles. The concern is whether to consider GIS a tool, a toolmaker, or a science (Heywood, 1990; Wright et al., 1997a, b; Pickles, 1997). Such arguments are of little importance here except to provide readers with an awareness of ongoing philosophical exchanges.

Health Services Research

Two recent sources provide consensus on the definition of health services research. The first definition is from the editors of *Health Services Research: An Anthology* and reads as follows: "The central feature of health services research is the study of the relationships among structures, processes, and outcomes in the provision of health services" (White et al., 1992, p. xix). The second definition is drawn from the mission statement of *Health Services Research*, the official journal of the Association of Health Services Research. The mission statement reads as follows: "[T]o enhance knowledge and understanding of the financing, organization, delivery, and outcomes of health services through publication of thoughtful, timely, rigorously conducted, state-of-the-art research and thinking" (Association of Health Services Research, 1997).

CONCERNS ABOUT USING GIS

GIS is no panacea! Matthews (1990) recommends that epidemiologists, medical geographers, and other spatial scientists adopt cautionary and critical approaches in using geographic information systems. First, there is no consensus on the definition of geographic information systems. Second, there is a potential for geospatial techniques to overshadow the research question(s). Third, GIS often suffer from the "4 in's" (intensive, inaccessible, inaccurate, and incomplete). That is, GIS often have intensive data requirements as well as problems of inaccessible, inaccurate, and incomplete data. Fourth, data might not be of appropriate geographic scale (aggregation) or temporal frames to support specific investigations in question (Albert et al., 1995; Heywood, 1990).

GEOGRAPHIC INFORMATION SYSTEMS AND HEALTH SERVICES RESEARCH

Our definitions of GIS and health services research constrain the selection of studies for this review. They exclude epidemiologic, geomedical, and public and environmental health studies that might have crossover or peripheral interest to health services research. Note, however, that literature reviews exist for these fields (see Chapter 1). Some 20 studies (found through searches of appropriate electronic databases using the terms GIS and variant spellings of GIS) were found to be appropriate for our purposes. Existing studies have focused on (1) physicians, (2) hospitals and other health care facilities, and (3) surveillance, monitoring, and emergency response. Physician studies have considered the location and distribution of physicians. Distance (straight line, relative, road, and time) is a common theme in studies of access to hospitals and other health care facilities. The other preoccupation involves determining hospital market areas (i.e., penetration rates) and hospital services areas. Finally, the locational and query dimensions of GIS are ideal for a wide variety of surveillance, monitoring, and emergency response activities.

PHYSICIAN DISTRIBUTIONS

One of the earliest advocates of the potential of using geographic information systems to monitor the distribution of physicians was Jacoby (1991). Nevertheless, just a handful of studies have used GIS to analyze geographic aspects of physician databases (Albert, 1995; Prabhu, 1995). Even fewer studies exist that examine the geography of physician assistants, nurses, chiropractors, and other practitioners within the health care system; the geography of these non-physician practitioners should also be explored.

Professional and personal criteria often guide physicians in selecting practice locations. Jankowski and Ewart (1996) developed a prototype spatial decision support system (SDSS) that allows physicians to select and evaluate potential locations for rural health practices. This SDSS integrated a GIS database (ARCINFO), a map visualization module (ArcView), and a multicriteria evaluation model (MCE). The SDSS was demonstrated using data for Idaho. Information representing aspects of (1) professional criteria: population, population density, primary care physicians needed, primary care need ranking, percent working in health care, primary care service areas ranked by primary care need, places without health care professionals, hospitals, rural health clinics, primary care service areas approved for loan repayment program funds, percent receiving Medicare and Medicaid, and fertility rate; and (2) personal criteria: major industry, tourism rank, percent unemployment, percent below poverty level, percent minorities, percent with college degree, urban center (>10,000 population), colleges, commercial airports, Amtrak stops, and alpine ski resorts were all brought under the domain of the SDSS. ArcView provided the software to visualize these themes (professional and personal criteria); for example, the themes of rural health clinics, places recruiting health professionals, and primary care service areas were integrated into one view (map).

Health professionals exploring potential practice locations might interact with the SDSS using one or more methods. First, one might select a point symbol (symbol for places, hospitals, etc.) from the scrollable pop-up window (map). The professional and personal criteria associated with the selected point(s) are retrieved and presented in tabular form for user inspection. Second, the user might search the database using a logical expression, such as (Medicare <20) *and* (docs ≥1.0). Those records meeting the criteria (i.e., records indicating less than 20% of population in Medicare and physician requirements one or greater) are highlighted in the database and output as a map as well. Third, the user can select up to eight criteria (from among 24) and indicate whether each represents a benefit (more is better) or a cost (less is better). Next the user assigns criteria priorities (i.e., indicate if the first criterion is more important or of equal importance to the second criterion) for a pairwise comparison. The MCE model evaluates and ranks locations and provides health practitioners some rationale for selecting among potential practice locations.

In another application, physician licensure data from the North Carolina Board of Medical Examiners were input within a geographic information system to examine location characteristics and distribution patterns of physicians with multiple medical practices (Albert, 1995, 1997b; Albert and Gesler, 1996, 1997). The

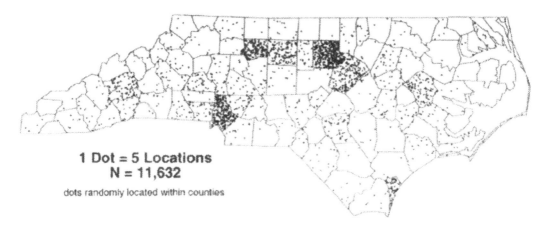

Figure 4.1. Primary Locations of Medical Practice, North Carolina, 1992. *Source: The North Carolina Geographer, 5,* D.P. Albert and W.M. Gesler. Comparing Physicians' Primary, Secondary, and Tertiary Practices Using Geographic Concepts: North Carolina, 1992, pp. 41–51, 1996. Reprinted with permission from the North Carolina Geographical Society.

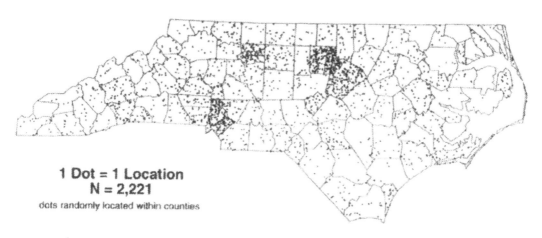

Figure 4.2. Secondary and Tertiary Locations of Medical Practice, North Carolina, 1992. *Source: The North Carolina Geographer, 5,* D.P. Albert and W.M. Gesler. Comparing Physicians' Primary, Secondary, and Tertiary Practices Using Geographic Concepts: North Carolina, 1992, pp. 41–51, 1996. Reprinted with permission from the North Carolina Geographical Society.

physician database included the following fields: physicians' gender, race, specialty, and the hours per week in medicine, setting, ZIP Code, city, county, and state for physicians' primary (Figure 4.1), secondary, and tertiary practice locations (Figure 4.2). Other information (metropolitan status, shortage status, settlement size) was joined with the original physician database via common data fields.

Some of the results of the multiple practice study are summarized below. The foci of the 2,221 secondary and tertiary practice locations were within metropolitan counties, especially those with medical schools and or large medical complexes. Physicians' secondary and tertiary locations were outside the county of physicians' primary locations (62%); however, these secondary and tertiary locations were more often in similar counties with respect to metropolitan status and shortage status. Secondary and tertiary locations were found further down the urban hierarchy than primary locations. For example 57% of secondary and 65% of tertiary locations, compared with 35% of primary locations, were in settlements with populations under 20,000. Twenty-four miles separated primary from secondary locations, 30 miles separated primary and tertiary locations, and 35 miles separated secondary and tertiary locations. One scenario suggests physicians were targeting secondary and tertiary locations just outside the immediate area of primary locations to service adjacent areas, perhaps suburban realms of core metropolitan areas.

Some communities have no or an inadequate number of physicians serving their population. Such communities might use a GIS to find physicians practicing within the region for potential referrals or recruitment (Prabhu, 1995; Albert, 1996). To illustrate, databases including information on physicians' practice locations (ZIP Code, city, and county), physicians' characteristics (e.g., physician specialties), and latitude and longitude coordinates of municipalities were input to a GIS environment (Albert, 1996). Next, a 45-mile radius was drawn about a rural community to search for physician practices. For example, one search found over 200 physicians with secondary and tertiary practices within a 45-mile radius of Snow Hill, North Carolina, in 1992. Physicians within the region might use these data to refer patients to other neighboring physicians. Clinics or other health care facilities might use this information to target, for recruitment purposes, physicians within some fixed radius.

HOSPITALS AND OTHER HEALTH CARE FACILITIES

Distance Measures

Distance to health care facilities is an important factor for patients, practitioners, and administrators. Some of the most common functions of geographic information systems are its measurement functions. The citations given below illustrate straight-line (Kohli et al., 1995; Love and Lindquist, 1995), relative (Lee, 1996), time (Furbee and Spencer, 1993), and road or network distance (Walsh et al., 1997). Of course, the selection of an appropriate distance measure depends on the specifics surrounding a particular investigation. For example, Phibbs and Luft (1995) found the correlation between travel time and straight-line distance between patients' residences and nearest hospitals was 0.987 for all observations and 0.826 for distances less than 15 miles. However, these authors noted exceptions to these correlations for studies focusing on specific hospitals, very small hospitals, or hospitals located in dense urban areas. One must decide whether the more elaborate measures of distance (i.e., time or network) are worth the additional effort and expense.

Population and property registries provide data to calculate distance to care. For example, in Sweden citizens are required to notify the population registry of address changes. Thus records from the population registry can be linked with a property registry using an address field common to both databases. Since the property databases includes the x- and y- coordinates of the center of each property (address), the straight-line distance to each person's assigned primary health center can be determined. Kohli et al. (1995) calculated the number of persons plus the average, maximum, and minimum distance to primary health center by sex and age (0–6, 7–64, 65–74, and >75). Further, population data were aggregated to commune level for which minimum, maximum, median, and average distances to primary health center were determined. For each commune, the number and percent of population falling within various distance bands (i.e., 0–1000 m, 1–5 km, 5–10 km, >10 km) to primary health centers were calculated. In Sweden, as in most other Nordic countries, excellent registries provide the requisite data for a GIS. This allows researchers and others exploring Nordic health services to come to grips with distance-to-care issues better than anywhere.

For another example of the straight-line method, Love and Lindquist (1995) used a GIS to measure distance of aged populations to hospitals in Illinois. They generated isarithmic maps showing four distance contours (0 to 5, >5 to 10, >10 to 15, >15 to 20, >20 miles) from the closest hospital and the closest geriatric hospital. Eighty percent of the Illinois aged population was within 4.8 miles of one hospital and 11.6 miles of two. There were substantial differences in distances to first, second, third, fourth, and fifth closest hospitals for the aged population living within and outside metropolitan statistical areas. However, the authors found no evidence that access or distance is different for the aged than for the general population.

Lee assessed the relative distance of San Francisco's homeless population to clinics and hospitals. Using MapInfo, he created a map showing clinic locations relative to the distribution of the homeless population. Using a one-mile buffer around hospitals with emergency rooms, Lee noted that most homeless shelters and free food were within these zones. Similarly, a 0.75-mile buffer around clinics also contained most of the homeless shelters and free food. The author suggested that the homeless were geographically accessible to health care; however, he realized that a "lack of insurance excludes the majority of the homeless from most traditional fee-for-service health care providers, health maintenance organizations, and hospitals other than for emergency care" (Lee 1996, p. 46). Here, with such a transient population, it would be difficult to determine straight-line distances between homeless populations and clinics and emergency rooms. The solution is to examine the relative location or juxtaposition of the major nodal points of the homeless (i.e., the shelters and free food) to that of the health care facilities providing care to the homeless.

In another study (Furbee and Spencer, 1993), travel times for a county's residential population to the county's only hospital were determined using a raster-based GIS (raster GIS stores map overlays as cells in a row and column matrix). The transportation network of the county was brought into the GIS as five separate layers. The five overlays included primary highways, secondary highways, county roads, neighborhood roads, and jeep trails. It was assumed that vehicular speed would average 50 miles per hour on primary roads, 40 mph on secondary roads, 30 mph on county roads, 20 mph on neighborhood roads, and 10 mph on jeep trails. Each of the five transportation overlays was reclassified to reflect the number of seconds required to traverse a cell (each cell was 296 ft × 296 ft). The five transportation overlays were then combined to produce a single layer that represented the friction (in seconds) of traversing each cell (296 × 296). If two cells from different overlays (i.e., primary and secondary roads) intersected, the higher frictional value was assigned to the composite layer. Next, the row and column address (i.e., the cell) corresponding to the hospital location was determined and this also constituted a map layer. The overlay composite showing frictional values and the overlay indicating the hospital location was used to calculate the time (in seconds) it took to travel from each cell to the hospital. These travel times were classified into four bands of ten minutes each. The 10-minute travel time bands were overlain onto census boundary files in order to estimate the population residing in each band. The authors then tested their model by driving to the hospital, at the posted speed limit, from two of the furthest locations in the county. From this test they recorded travel times from 51 points along the route. All of these 51 points fell within the 10-minute time bands previously established.

Another variation of distance measurement incorporates road and time distance. Walsh et al. (1997) assigned patients to the nearest available hospitals with remaining capacity (beds). This required patient discharge data to determine patients' residential location, U.S. TIGER/Line files of the transportation network, and latitude and longitude coordinates of hospital locations. Patients were assigned to nearest hospitals, based on time along transportation network from residence to hospital, using the ARC/INFO Network module. These data were then used to construct normative hospital service areas (see next section for elaboration).

Market and Hospital Service Areas

Geographic information systems have been used in siting hospitals (Marks et al., 1992), to delineate market areas, and to construct hospital service areas. The following examples illustrate the construction of a market share map at a fine geographic scale (i.e., the census tract), delimiting hospital service areas using Thiessen polygons, and constructing hospital service areas by integrating patient data, the transportation network, and hospital supply (i.e., hospital beds) using a network analysis.

St. Mary's Medical Center in Duluth, Minnesota used a geographic information system to map its market share. Rather than just aggregating patient data to the 10 ZIP Codes within Duluth, a more detailed geographic scale, the census tract, was used (Miller, 1994). Patient residences were address-matched to MapInfo's StreetInfo file. Next, a boundary file of the 44 census tracts of Duluth was placed over (overlay) the address-matched patient origin data. Then the number of residences contained within each census tract was summed. Market penetration rates could then be determined by dividing the number of patients into total population for each census tract. These penetration rates, per census tract, were output as a market share map. Thus, GIS provided St. Mary's Medical Center the capability of enhancing existing locational data available from patient origin records.

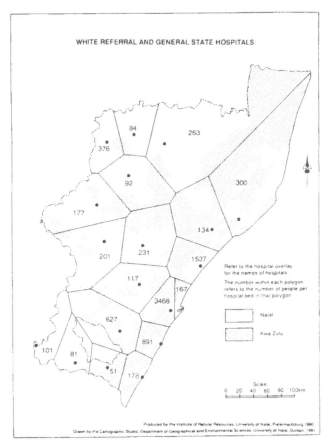

Figure 4.3. White Referral and General State Hospitals. *Source: South African Medical Journal,* 80, M. Zwarenstein, D. Krige, and B. Wolff. The Use of Geographic Information System for Hospital Catchment Area Research, pp. 497–500, 1991. Reprinted with permission.

While the original patient data included street addresses and ZIP Codes the enhanced database provided census tract information. This allowed for a more detailed areal geographic view of the Medical Center's market share.

Zwarenstein et al. (1991) calculated changing person-to-bed ratios within hospital catchment areas in Natal/KwaZulu, South Africa, in light of removing race restrictions on admissions (Figures 4.3, 4.4, and 4.5). The hospital catchment areas were estimated using Thiessen polygons, a neighborhood operation performed within GIS. Thiessen polygons delineate boundaries around nodes (hospitals) to create catchment areas. The basic premise is that the boundaries between catchment areas are equidistant from nodal centers (hospitals). Using a GIS to construct Thiessen polygons to represent service areas is an alternative to the more usual method of aggregating geopolitical or administrative boundaries (i.e., counties, townships, etc.). Zwarenstein et al. (1991) found that one-half of the catchment areas for blacks had person/bed ratios of above 275, contrasting with one-third of catchment areas for whites with person/bed ratios of above 275. The unequal

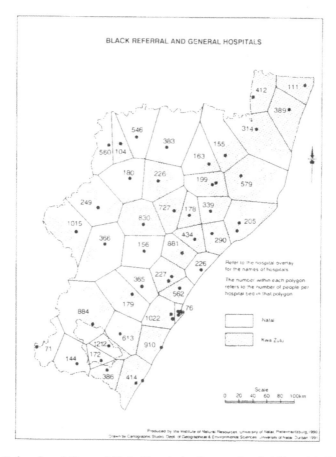

Figure 4.4. Black Referral and General State Hospitals. *Source: South African Medical Journal*, 80, M. Zwarenstein, D. Krige, and B. Wolff. The Use of Geographic Information System for Hospital Catchment Area Research, pp. 497–500, 1991. Reprinted with permission.

distribution of resources is attributable to a shortage of hospitals in the remote rural areas. Increasing access to clinics in the rural areas might relieve congestion at the general and referral hospitals and reduce person/bed ratios.

Walsh et al. (1997) constructed normative hospital service areas for a 16-county region around Charlotte and Mecklenburg County, North Carolina. They used location/allocation modeling to optimize travel time between patients and 25 hospitals. Their objectives were to conduct a network analysis that integrated patient, transportation, and hospital characteristics. Patient discharge data, location of hospitals, and TIGER/Line files (digital files of the transportation network), and census variables were brought within a GIS database. Normative service areas were defined by allocating patients to hospitals while minimizing travel time and factoring in hospital supply (beds). Travel times were established by assigning estimated average speeds for road types (72 kph for primary roads and highways, 64 kph for secondary roads, 30 kph for connecting and county roads, and 15 kph for neighborhood and city streets). Normative service areas were defined in this

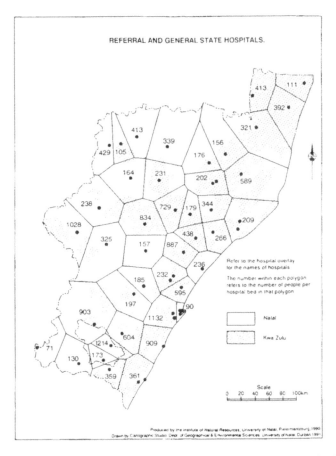

Figure 4.5. Referral and General State Hospitals. *Source: South African Medical Journal,* 80, M. Zwarenstein, D. Krige, and B. Wolff. The Use of Geographic Information System for Hospital Catchment Area Research, pp. 497–500, 1991. Reprinted with permission.

manner for total patient discharges using 1991 patient data (Figure 4.6) and again assuming a doubling of patient demand while maintaining supply elements constant (Figure 4.7). In a similar fashion normative service areas were defined using DRG 391 (DRG = diagnosis-related group; 391 = Normal Newborn) with 1991 data and again assuming the scenario where demand doubles. With total discharge data the hospital service areas constricted considerably with a doubling of demand. Using just DRG 391 only minor changes in the areal extent were noted with a doubling of demand. Perhaps this suggests an oversupply of resources currently devoted to DRG 391. Total discharges and DRG 391 were chosen for illustrative purposes for using the network analysis to create normative hospital service areas. The authors note the potential of constructing such normative service areas for other DRGs. Their results suggest that network analysis "is an effective approach for exploring a variety of healthcare scenarios where changes in the supply, demand and impedance variables can be examined within a spatial context" (Walsh et al., 1997, p. 244).

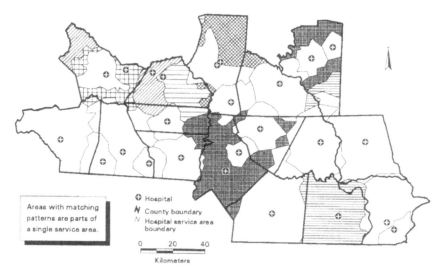

Figure 4.6. 1991 Patient Population, Network Analysis Service Areas. *Source:* Reprinted with permission of the Health Research and Educational Trust, copyright 1997.

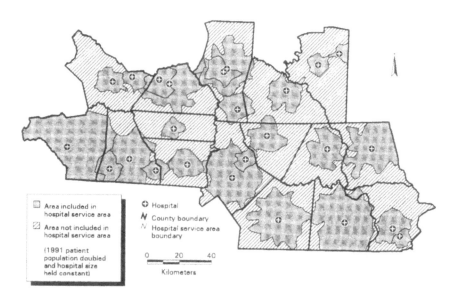

Figure 4.7. 1991 Patient Population x 2, Network Analysis Service Areas. *Source:* Reprinted with permission of the Health Research and Educational Trust, copyright 1997.

MONITORING, SURVEILLANCE, AND PLANNING

GIS have been developed for surveillance and outreach programs. Only those programs that were geared to provide services or assess outcomes are reviewed here. Certainly, GIS have a tremendous potential for monitoring and surveillance programs; however, these are more within the realm of environmental and public health than in health services research. (Please note that some portions of this section were repeated or expanded from Chapter 3 for the convenience of the reader.)

Solarsh and Dammann (1992) put together a community paediatric information system (CPIS) for the Edendale Health Ward in Southern Natal, South Africa. Inpatient records, from Edendale Hospital (the main referral hospital) formed the basis for creating a CPIS. The objectives of the CPIS were to provide descriptive data on preventable childhood diseases, to describe incidence rates along a hierarchical order of geographical units (Edendale Health Ward, magisterial district, and subdistrict), and to monitor longitudinal trends in child health. The CPIS consisted of dBase III Plus, a software spreadsheet application for input and editing of hospital records, and Epi Info, an epidemiological program from the Centers for Disease Control. Inpatient records were coded on the following criteria: date of admission, duration of stay, sex, age, diagnostic code, address (5 categories—Edendale Health Ward, magisterial unit, subdistrict, nearest clinic, and nearest school), outcome, notification, vaccination status, and others. Specifically, this CPIS was used to examine measles admissions from 1987 to 1990 (four full years). Monthly admission rates for measles were shown (using histograms) to decline dramatically from 1987 (year of epidemic) to 1988 with continuing gradual declines in 1989 and 1990. Data on measles were presented at various geographic scales including inside vs. outside the Edendale Health Ward, magisterial district (Figure 4.8), and subdistrict. The inside vs. outside data revealed that a decreasing proportion of cases was coming from the Edendale Health Ward as opposed to the surrounding health authorities. Again, examination of data at the magisterial district and subdistrict revealed a decline in the incidence of measles from epidemic year (1987) to 1990; however, specific districts and subdistricts stand out as having substantially higher rates than other districts and subdistricts. The CPIS can be used to delimit susceptible (unvaccinated) and unsusceptible populations at various geographic scales. This would allow health care workers to respond in a more focused manner when outbreaks occur. Immunization programs might also be evaluated using data within the CPIS to correlate vaccination rates with measles incidence rates.

Data from the Arizona State Immunization System (ASIS) were integrated with a GIS software package (ArcView) to provide cartographic output that proved useful in monitoring immunization status and disease occurrence (Popovich and Tatham, 1997). One specific application involved a geographical analysis of a pertussis epidemic occurring in Arizona along the California and Mexican borders. The cartographic capabilities of a GIS were useful in creating a series of maps depicting immunization status and cases at the census block and street level scales for Yuma County, Arizona. For example, one figure showed the locations of immunized children under age 2 and pertussis cases as overlays to a thematic shading map of the total number of children under age 2 per census. Another figure showed the residential locations of children with fewer than three DTP shots and pertussis cases with respect to two-kilometer radii drawn around each of the two clinics in Yuma County. Detailed data from central registries (i.e., ASIS) contained geographic information useful in automating mapping to specific locales.

The Youth Environment Study (YES), a nonprofit organization, used MapInfo to stem the diffusion of HIV/AIDS among intravenous drug users in San Fran-

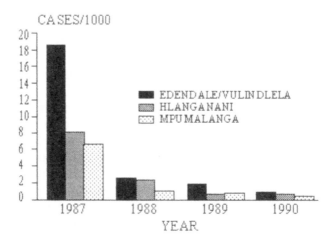

Figure 4.8. Measles Incidence Rates by Magisterial District, 1987–1990. *Source: South African Medical Journal,* 82, G.C. Solarsh and D.F. Dammann. A Community Paediatric Information System: A Tool for Measles Surveillance in a Fragmented Health Ward, pp. 114–118, 1992. Reprinted with permission.

cisco (Fost, 1990). YES field workers distributed bleach for needle sterilization and condoms in efforts to provide barriers to the transmission of HIV among this high-risk population. Field workers also solicited information from the intravenous drug users via a questionnaire. The questionnaire asked for demographic data such as race and ethnicity, behavioral data such as needle sharing habits (e.g., Do IV users share their needles? If so, what percent of needles are cleaned prior to reuse?), and locational data such as the street intersections which drug users are known to frequent. These data were input into the desktop mapping package and address-matched to a street file. Field workers could then be matched with an IV user of the same racial or ethnic background and other characteristics by searching the GIS database per user-defined logical expressions. Field workers were sent into the city with a street-level map showing the locations of IV users who shared unclean needles. In this manner personnel (field workers) and reserves (bleach and condoms) resources were used to maximum advantage.

There have been other applications of geographical information systems for emergency response planning and operations (Furbee, 1995; Coleman, 1994). GIS have been used to locate the nearest rescue vehicles to an accident (Van Creveld, 1991) and to predict ambulance response times to accident locations (Jones, 1993). Dunn and Newton (1992) discuss optimizing routes for emergency planning applications within a GIS.

DISCUSSION

Returning to our initial definition of health services research we find that current applications of GIS focus on structural or delivery aspects rather than processes and outcomes. This is understandable given that these elements lend themselves to spatial representation (i.e., the distribution of providers and facilities). Whether GIS can assist in understanding process and outcome elements of health services research is unknown. Folding GIS into all components (structures, process, and outcomes) of health services research might increase its standing within the discipline. Currently, GIS occupies a very peripheral position in health services research.

So what has been the contribution of GIS to health services research? In terms of the number of published articles in professional and academic journals/magazines the output is small. Of course this represents just the tip of the iceberg. There is no knowing the extent of GIS usage in government agencies and health care industries, although software venders are vigorously targeting these sectors as potential customers. GIS usage also surfaces in conference proceedings and this probably represents a sizeable volume of research; however, the more original paper, from such proceedings often find placement in academic journals at some later date. GIS is still in the expansion phase; early adopters are exploring the potential of GIS. The number of applications should increase exponentially in the next several years as the diffusion process continues. At this time it is difficult to say with any confidence whether or not GIS has contributed to the improvement of health, health services, or our understanding of disease. Those uses of GIS involving the real or near time analysis of immunization status appear to be the most dramatic examples of where the potential to intervene could translate into slowing an epidemic (Solarsh and Dammann, 1992; Popvich and Tatham, 1997).

Given the small number of studies using GIS in health services research, to what extent have the potential functions of GIS been utilized? The classification of GIS functions into four groups: (1) maintenance and analysis of the spatial data, (2) maintenance and analysis of the attribute data, (3) integrated analysis of the spatial and attribute data, and (4) cartographic output formatting providing an outline to assess whether researchers are taking advantage of the full power of GIS. Obviously, research should not be technique driven; however, one ought to know the technical arsenal available to investigators.

Maintenance and Analysis of Spatial Data and Attribute Data

Inherent within GIS is the capability for the maintenance and analysis of spatial (group 1) and attribute data (group 2). Note that most studies reviewed contained one or more layers of geographic files. For example, Love and Lindquist (1995) include boundary files for a state, census blocks, and hospital locations in their assessment of access of the aged population to hospitals. Common also were studies that have taken advantage of the attribute maintenance and editing functions of GIS. Note, for example, that Albert (1997b) linked data on settlement size, metropolitan status, and health professional shortage status to the physician database via common fields to produce an enhanced database. Group 1 and 2 functions provide the prerequisite spatial and attribute data for group 3 (integrated analysis of spatial and attribute data) and group 4 (cartographic output formatting) functions.

Integrated Analysis of Spatial and Attribute Data

The more powerful GIS capabilities exist within group 3, integrated analysis of spatial and attributed data. Remember, group 3 consists of four subgroups of functions that include retrieval/classification/measurement, overlay operations, neighborhood operations, and connectivity functions. Most studies use the retrieval, classification, or measurement functions of a GIS (Fost, 1990; Solarsh and Dammann, 1992; Furbee and Spencer, 1993; Love and Lindquist, 1995; Jankowski and Ewart, 1996; Albert 1997b; Popovich and Tatham, 1997). Surprisingly, just one study used an overlay operation (Furbee and Spencer, 1993). Here, an overlay operation refers to mathematical (addition, subtraction, multiplication, and division) or logical (if-then statements) combinations of data layers.

Aronoff (1989, p. 211) states that "[N]eighbourhood operations evaluate the characteristic of the area surrounding a specific location." Search, line-in-polygon and point-in-polygon, topographic functions, Thiessen polygons, interpolation, and contour generation are some forms of neighborhood operations. Just three of the six neighborhood operations have found usage in health services research; these include point-in-polygon, Thiessen polygons, and contour generation. Search functions canvass spatial databases to target for features (e.g., hospitals) and assign a value (e.g., average, diversity, majority, maximum, minimum, total) that describes associated characteristics (e.g., population, physicians, other facilities) occurring within some distance (or occurring within some geometric window such as a circle). No use of the search function is evident in the health services research literature; however, search functions have found use in epidemiological studies investigating cancer clusters (Openshaw et al., 1987). No uses of line-in-polygon (aggregating lines occurring within polygons) were found in the literature. Perhaps line-in-polygon functions might be useful to analyze vectors depicting flows of patients to physicians, physician referrals, and hospitals. Point-in-polygon is a method which counts points contained within a polygon; this is useful to aggregate point data into some existing or other user defined areal units. Point-in-polygon functions were used to aggregate patient residences into census tracts (Miller, 1994) and communes (Kohli et al., 1995).

Topographic functions calculate values that describe surfaces (i.e., elevation, slope, and aspect). Bashshur et al. (1970) represented population, physician, and hospital data as 3-D statistical surfaces in the 1970s. No recent attempts to create similar surfaces were found in recent HSR literature. 3-D surfaces provide a spectacular way to visualize data and health services research might reconsider the use of topographic functions. Interpolation functions predict unknown values from surrounding known values. No examples of interpolation exist in health services research; however, this function is useful in epidemiological studies to predict the diffusion of diseases (Eddy and Mockus, 1994).

Thiessen polygons can produce boundaries equidistant from a set of nodal points to construct a regionalization. Noteworthy is Zwarenstein et al.'s (1991) use of Thiessen polygons to define hospital service areas and measure bed-to-population ratios before and after elimination of race restrictions on hospital admissions. Love and Lindquist (1995) generated contour lines (5-mile increments) representing the distance to the nearest hospital with geriatric facilities for persons age 65 and over; Challender and Root (1994) describe isochrones (contours of equal time) representing access to rural health services.

The last subgroup of neighborhood operations consists of connectivity functions. These functions accumulate values over the network or area being traversed. Aronoff (1989, p. 220) lists and describes contiguity measures, proximity, network, spread, seek, intervisibility, illumination, and perspective view as neighborhood operations. Most of the functions in this subgroup have not found usage in health service, research, although some of these unused functions might have applications. For example, given a set of criteria, contiguity measures might be used to regionalize health services; spread functions might be able to suggest the direction(s) to target immunization programs; and seek functions might be able to find suitable sites for health care facilities. Intervisibility, illumination, and perspective view functions find most application in the earth sciences to present maps and graphs from different vantage points.

Of the remaining functions, only proximity (buffers) and network analyses have found applications in health services research. For example, Albert (1996) retrieved records from a physician database of office locations within a 45-mile radius of a town. Lee (1996) found that 1-mile and 0.75-mile buffers around emergency rooms and clinics contained most of the homeless shelters and free food (finding excellent geographic access while recognizing poor financial access). Walsh et al. (1997) integrated patient, transportation, and hospital characteristics within a network analysis to produce normative hospital service areas. Other applications of network analysis or routing are found in the emergency planning literature (Van Creveld, 1991; Jones, 1993; Dunn and Newton, 1992).

Output Formatting (Cartographic)

One expects studies using GIS to include an extensive use of maps. However, a number of studies using the retrieval and measurement functions of GIS present information in tabular rather than cartographic form. Some examples of effective cartographic output (maps) include contours to depict distance to the nearest hospital (Love and Lindquist, 1995); Thiessen polygons to create hospital service areas; network analysis to construct hospital service areas (Walsh et al., 1997); 10-minute travel bands around a county hospital to correspond with travel time (Furbee and Spencer, 1993); proportional symbols to depict patient origins (Gordon and Womersley, 1997); buffers to present the juxtaposition between homeless sites (shelters and free food) and emergency hospitals and clinics; and point symbols to represent immunological status of children. The opportunities for displaying the informative results of analysis using GIS are virtually unlimited.

CONCLUSIONS

GIS technologies are just beginning to diffuse into the realm of health services research. Only around two dozen papers exist (excluding proceedings and reports) that integrate health services research with a geographical information system.

These papers fall under the subject headings of physicians, hospitals, and monitoring, surveillance, and emergency response. The full functional potential that GIS technologies offer has not been realized. While most studies have employed GIS for retrieval, classification, and measurement, few studies have used some of the more advanced functions. Granted, research should not be driven by technology; however, the full potential of GIS is being underutilized. There are numerous situations in health services research where some of the more advanced GIS functions could improve research.

REFERENCES

Albert, D. 1994. Geographic information systems (GIS). In *Geographic Methods for Health Services Research: A Focus on the Rural-Urban Continuum,* T.C. Ricketts, L.A. Savitz, W.M. Gesler, and D.N. Osborne (Eds.), pp. 201–206. Lanham, MD: University Press of America.

Albert, D. 1995. Is there a doctor near the house? MapInfo analyzes health care access in North Carolina. *GlobalNews* (Summer):7.

Albert, D. 1996. Dimensions of multiple locations of medical practice: North Carolina, *1992.* Unpublished Ph.D. Dissertation. University of North Carolina at Chapel Hill.

Albert, D. 1997a. Synopsis and bibliographic resource of medical-GIS applications. In *The National Center for Geographic Information and Analysis: GIS Core Curriculum for Technical Programs,* M. Goodchild et al. (Eds.), Santa Barbara: University of California. (hhttp://www.ncgia.ucsb.edu/education/curricula/cctp/applications/med_bibliography.html).

Albert, D. 1997b. Monitoring physician locations with GIS. In *The National Center for Geographic Information and Analysis: GIS Core Curriculum for Technical Programs,* M. Goodchild et al. (Eds.), Santa Barbara: University of California. (hhttp://www.ncgia.ucsb.edu/education/curricula/cctp/applications/albert.html).

Albert, D. and W.M. Gesler. 1996. Comparing physicians' primary, secondary, and tertiary practices using geographic concepts: North Carolina, 1992. *North Carolina Geographer* 5:41–51.

Albert, D. and W.M. Gesler. 1997. Multiple locations of medical practice in North Carolina: Findings and health care policy implications. *Carolina Health Services and Policy Review* 4:55–75.

Albert, D.P., W.M. Gesler, and P.S. Wittie. 1995. Geographic information systems and health: An educational resource. *Journal of Geography* 94(2):350–356.

Antenucci, J.C., K. Brown, P.L. Croswell, M.J. Kevany, and H. Archer. 1991. *Geographic Information Systems: A Guide to the Technology.* New York: Van Nostrand Reinhold.

Aronoff, S. 1989. *Geographic information systems: A Management Perspective.* Ottawa: WDL Publications.

Association of Health Services Research. 1997. Mission Statement. *Health Services Research* 32(1):xix.

Barnes, S. and A. Peck. 1994. Mapping the future of health care: GIS applications in health care analysis. *Geo Info Systems* 4(4):30–39.

Bashshur, R.L., G.W. Shannon, and C.A. Metzner. 1970. The application of three-dimensional analogue models to the distribution of medical care facilities. *Medical Care* 8(5):395–407.

Briggs, D.J. and P. Elliot. 1995. The use of geographical information systems in studies on environment and health. *World Health Statistics Quarterly* 48(2):85–94.

Challender, S. and J. Root. 1994. Isochrones: a method of tracking and analyzing geographic access to rural health care. *Geo Info Systems* 4(6):33–34.

Coleman, D. 1994. GIS Canada: Road network partnerships paying off. *GIS World* 7(2):30.

Cowen, D.J. 1990. GIS versus CAD versus DBMS: What are the differences? In *Introductory Readings in Geographic Information Systems*, D.J. Peuquet and D.F. Marble (Eds.), pp. 52–61. London: Taylor & Francis.

Croner, C.M., J.S. Sperling, and F.R. Broome. 1996. Geographic information systems (GIS): New perspectives in understanding human health and environmental relationships. *Statistics in Medicine* 15 (17–18):1961–1977.

Dunn, C.E. and D. Newton. 1992. Optimal routes in GIS and emergency planning applications. *Area* (3)24:259–269.

Eddy, W.F. and A. Mockus. 1994. An example of the estimation and display of a smoothly varying function of time and space: The incidence of the disease mumps. *The Journal of the American Society for Information Science* 45(9):686–683.

Furbee, P.M. 1995. GIS in Raleigh County: Small towns with a big database. *Journal of Emergency Medical Services* 20(6):77, 79, 81.

Furbee, P.M. and J. Spencer. 1993. Using GIS to determine travel times to hospitals. *Geo Info Systems* (September):30–31.

Fost, D. 1990. Using maps to tackle AIDS. *American Demographics* 12(4):22.

Glass, G.E., J.L. Aron, J.H. Ellis, and S.S. Yoon. 1993. *Applications of GIS Technology to Disease Control*. Baltimore: The Johns Hopkins University, School of Hygiene and Public Health, Department of Population Dynamics.

Gordon, A. and J. Womersley. 1997. The use of mapping in public health and planning health services. *Journal of Public Health in Medicine* 19(2):139–147.

Heywood, I. 1990. Geographic information systems in social sciences. *Environment and Planning A* 22(7):849–854.

Jacoby, I. 1991. Geographic distribution of physician manpower: The GMENAC legacy. *Journal of Rural Health* 7(4 Suppl):427–426.

Jankowski, P. and G. Ewart. 1996. Spatial decision support for health practitioners: Selecting a location for rural health practice. *Geographical Systems* 3:279–299.

Jones, A. 1993. Using GIS to link road accident outcomes with health service accessibility. *Mapping Awareness & GIS in Europe* 7(8):33–37.

Kohli, S., K. Sahlen, A. Sivertun, O. Lofman, E. Trell, and O. Wigertz. 1995. Distance from the primary health center: A GIS method to study geographical access to health care. *Journal of Medical Systems* 19(6):425–436.

Lee, H. 1996. Health care for San Francisco's homeless. *Geo Info Systems* 6(6):46–47.

Love, D. and P. Lindquist. 1995. The geographical accessibility of hospitals to the aged: A geographic information systems analysis within Illinois. *Health Services Research* 29(6):629–651.

Marks, A.P., G.I. Thrall, and M. Arno. 1992. Siting hospitals to provide cost-effective health care. *Geo Info Systems* 2(8):58–66.

Matthews, S.A. 1990. Epidemiology using a GIS: The need for caution. *Computer, Environment and Urban Systems* 14(3):213–221.

Miller, P. 1994. Medical Center uses desktop mapping to cut costs and improve efficiency. *Geo Info Systems* 4(4):40–41.

Mott, K.E., I. Nuttall, P. Desjeux, and P. Cattand. 1995. New geographical approaches to control of some parasitic zoonoses. *Bulletin of the World Health Organization* 73(2):247–257.

Nicol, J. 1991. Geographic information systems within the National Health Service: The scope of implementation. *Planning Outlook* 34(1):37–42.

Openshaw, S., M. Charlton, C. Wymer, and A. Craft. 1987. A Mark 1 geographical analysis machine for the automated analysis of point data sets. *International Journal of Geographical Information Systems* 1(4):335–358.

Pan American Health Organization. 1996. Use of geographic information systems in epidemiology (GIS-Epi). *Epidemiological Bulletin* 17(1):1–6.

Parr, D.M. 1991. *Introduction to Geographic Information Systems Workshop.* Wilmington, NC: The Urban and Regional Information Systems Association.

Phibbs, C.S. and H.S. Luft. 1995. Correlation of travel time on roads versus straight line distance. *Medical Care Research and Review* 52(4):532–542.

Pickles, J. 1997 Tool or science? GIS, technoscience, and the theoretical turn. *Annals of the Association of American Geographers* 87(2):363–372.

Public Technology, Inc. Urban Consortium for Technology Initiatives, and International City Management Association. 1991. *The Local Government Guide to Geographic Information Systems: Planning and Implementation.* Washington, DC: Public Technology Inc. and International City Management Association.

Popovich, M.L. and B. Tatham. 1997. Use of immunization data and automated mapping techniques to target public health outreach programs. *American Journal of Preventive Medicine* 13(2 Suppl):102–107.

Prabhu, S. 1995. A generalized framework for information systems for physician recruitment/referral. *Journal of Management in Medicine* 9(4):24–30.

Scholten, H.J. and M.J. de Lepper. 1991. The benefits of geographical information systems in public and environmental health. *World Health Statistics Quarterly* 44(3):160–170.

Solarsh, G.C. and D.F. Dammann. 1992. A community paediatric information system: A tool for measles surveillance in a fragmented health ward. *South African Medical Journal* 82(2):114–118.

Star, J. and J. Estes. 1990. *Geographic Information Systems: An Introduction.* Englewood Cliffs, NJ: Prentice Hall.

Taylor, D.R.F. (Ed.). 1991. *Geographic information systems: The Microcomputer and Modern Cartography.* Oxford, England: Pergamon Press.

Twigg, L. 1990. Health based geographical information systems: Their potential examined in the light of existing data sources. *Social Science and Medicine* 30(1):143–155.

Van Creveld, I. 1991. Geographic information systems for ambulance services: In *Geographic Information 1991: The yearbook for the Association of Geographic Information,* pp. 128–130. London: Taylor & Francis.

Walsh, S.J., P.H. Page, and W.M. Gesler. 1997. Normative models and healthcare planning: Network-based simulations within a geographic information system environment. *Health Services Research* 32(2):243–260.

White, K.L., J. Frenk, C. Ordonez, C. Paganini, and B. Starfield. 1992. *Health Services Research: An Anthology.* Washington, DC: Pan American Health Organization.

Wright, D.J., M.F. Goodchild, and J.D. Proctor. 1997a. GIS: Tool or science? Demystifying the persistent ambiguity of GIS as "Tool" versus "Science." *Annals of the Association of American Geographers* 87(2):346–362.

Wright, D.J., M.F. Goodchild, and J.D. Proctor. 1997b. GIS: Tool or science? Demystifying the persistent ambiguity of GIS as "Tool" versus "Science." *Annals of the Association of American Geographers* 87:373.

Wrigley, N. 1991. Market-based systems of health-care provision, the NHS Bill, and geographical information systems. *Environment and Planning A* 23(1):5–8.

Zwarenstein, M., D. Krige, and B. Wolff. 1991. The use of a geographic information system for hospital catchment area research in Natal/KwaZulu. *South African Medical Journal* 80(10):497–500.

Chapter Five

GIS-Aided Environmental Research: Prospects and Pitfalls

INTRODUCTION

Geographic information systems (GIS) technology increases the quality of information produced by environmental hazards and epidemiologic investigations by adding the dimension of context (Tim, 1995). GIS achieves its highest, best use in the information integration role, providing the infrastructure for combining the disparate types of information needed in environmental/ecologic studies (Nyerges et al., 1997). A wider range of complex disease-environment relationships can be unlocked by combining spatial analytic methods with GIS. The hope that this combination will help us realize a future free of human-induced cancer, leukemia and genetic mutation motivates many investigators. Fortunately, many of the causal and promotive variables involved in carcinogenesis are spatially distributed, and can be analyzed with spatial statistics.

Dimensions of Exposure Assessment

The accuracy of exposure estimates limits the success of both hazards and health outcome investigations. Conventional epidemiologic approaches aimed at assessing health effects resulting from exposure to hazardous waste sites have often failed to provide useful results due to three basic shortcomings: inadequate identification of the exposed population, lack of adequate health effect end points, and incomplete exposure measurements (Nuckols et al., 1994).

To assess exposure adequately, investigators need to know as much as possible about each of the following attributes of the problem (Moore, 1991; Stallones et al., 1992), including:

Substance Characteristics

- the toxicity, chemistry and physics of the substance(s) of interest (solubility/hygroscopicity, electrostasis, resuspendibility, and so forth),
- their release time periods, rates and concentration,

Transport

- the environmental media into which they are released (i.e., air, surface water, groundwater, soil surface or subsurface),
- how quickly the substances can move through those media,

Fate

- how their physical and chemical properties change under environmental conditions,
- how these changes affect the potential for harming population or environment,

Exposure

- the probability of uptake, taking into account the spatio-temporal distribution of the population with regard to the range and duration of exposure zones,
- the pathway into the body,
- where the substance moves once inside,
- what types of impact, and how much impact it is likely to have when it has completed its journey, and

Latency Period

- how long damage will take to manifest itself as a recognizable health effect.

Many studies are limited to a few of these aspects due to a lack of adequate modeling, data, computers and software, or other resources. Nonetheless, the more precise, complex and interdisciplinary the selected methods are, the more accurately they will model the ways that exposure effects are modified from release point to disease diagnosis (Sexton et al., 1992; Figure 5.1).

Advances in Understanding Environment-Health Relationships Using GIS

Proponents of the use of GIS have pointed out that its architecture is ideal for handling the complexities of a relatively large number of spatially distributed variables, and should emerge as a powerful tool in ecologic studies of exposure to environmental hazards and cancer etiology (Scholten and de Lepper, 1991; Stallones et al., 1992; Wartenberg, 1992; Croner et al., 1996). Nonetheless, the extent to which GIS-assisted ecologic studies can establish causation is vigorously debated (Waller, 1996a; Schwartz, 1994; Susser, 1994a, b). There is no doubt, how-

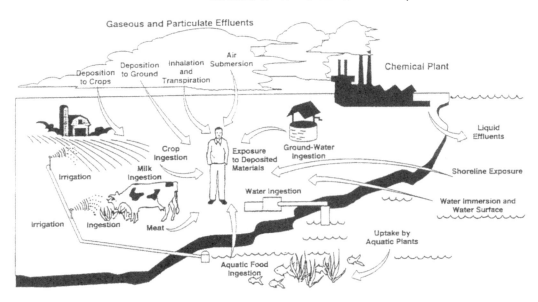

Figure 5.1. The ecologic approach considers all possible sources of exposure to contaminants, along with their pathways. *Source:* Sexton et al., 1992, p. 401.

ever, that these studies can be invaluable in placing disease in context, generating hypotheses, and justifying more expensive individual-level studies.

Two broad categories of environmental and health investigation are benefitting from the rapidly evolving capabilities of GIS-aided spatial analysis: estimating exposures from existing sources and modeling exposures anticipated from planned development. Current environmental monitoring methods have the potential to track ongoing exposures with a high degree of accuracy, limited only by the level of rigor dedicated to their deployment. Contemporary reconstructions of historical exposures such as those being performed at U.S. nuclear weapons production facilities can be more difficult, as there is a likelihood that they were either not measured or measured too crudely to be of much use. However, these efforts have become laboratories for developing some of the most advanced modeling techniques being performed (Shipler et al., 1996). Today, the planning stages of industrial, commercial and residential development usually involve an environmental impact assessment. Exposures projected by these studies entail higher levels of uncertainty as emissions are not available to reveal pathways and calibrate transport and deposition models. Without GIS technology, such assessments would be impossible to generate with any degree of timeliness and accuracy.

Software packages have evolved from unwieldy command-driven versions which once required specialist operators, to highly accessible graphical user interfaces. But with this increased accessibility come new opportunities to generate spurious associations instead of real answers. As GIS combines (individual-level data with) aggregate data collected at a wide range of resolutions, these ecologic studies have the potential to contribute information undiscoverable by any other means, but they also introduce opportunities for bias, inadvertent misrepresentation, or violation of confidentiality if used improperly (Waller, 1996a; Croner et al., 1996).

GIS Features Useful in Environmental Epidemiology

Numerous features of GIS are useful in environmental studies, including buffering, overlay and spatial query, nearest-neighbor identification, distance functions, interpolation, region-building, address-matching, and map production (Vine et al., 1997; Waller, 1996b; Twigg, 1990).

Buffering. When small differences in proximity to an exposure can result in important differences in its magnitude, the use of buffering can be helpful in distinguishing a population with high enough exposure to contrast with unexposed populations. Wartenberg and colleagues used a 100-meter buffer zone with block-level census data to characterize populations living very near high-voltage transmission lines (Wartenberg et al., 1993), whose exposure to electromagnetic fields rapidly weaken according to a logarithmic function of distance from the line.

Overlay and spatial query functions can augment individual-level or aggregate scale population data with spatially referenced attribute data. This permits the investigator to query the database for the spatial coincidence of features of interest. Overlay analysis was used to demonstrate GIS's usefulness in examining infant birthweights as indicators of environmental contamination. In this demonstration, maps of groundwater well locations, of an acquifer's extent and boundaries of contamination, and of residences around a toxic landfill were overlaid to generate a composite map showing which wells and residences were likely to be drawing from the acquifer's contaminated zone (Stallones et al., 1992).

Distance Calculations. The ability to make distance measurements enables a number of useful applications, such as calculating a household's distance from the nearest toxic waste site, determining effective response-time zones for providing emergency care services, or computing potentially harmful dimensions of a toxic plume or cloud (Croner et al., 1996). For example, Vine and colleagues (1997) used the distance-calculating function to supplement an individual-level database including blood pesticide levels with distance to nearby pesticide dump sites.

Nearest-neighbor analyses can exploit the distance between cases to perform a cluster analysis, or the presence of spatial autocorrelation in areal studies, to reveal whether contagious pathogens or environmentally borne carcinogens are associated with elevated incidence rates (Tobias et al., 1996; Glick, 1979). Some GIS packages such as Arc View and ARC/INFO can conduct nearest-neighbor analyses without exporting data. ARC/INFO and Arc View generate the values of "joins" (boundaries shared by two areal units) and "lags" (the number of areal units between the "source" unit and the "target" unit) required for representing spatial relationships in statistical models so that spatial autocorrelation can be quantified. Statistical models such as mixed models, multi-level models, and structural equation models can all handle autocorrelation when told which units are likely to carry it, but some users of these models fail to realize that spatial autocorrelation is not covered unless specific topological information, such as x-y coordinates or joins and lags, is included in the model.

PROBLEMS INHERENT IN GIS AND ECOLOGIC STUDIES

Three kinds of problems present special obstacles in GIS-enhanced ecologic studies: uncertain data quality, the risk of jeopardizing confidentiality, and difficulties in choosing a valid study design. The important features of each of these problem areas will be briefly reviewed here so that the reader can be alert to the ways these issues impact the study designs discussed below. Then various case studies representing topics of hazards management, environmental equity and comprehensive dose-to-effect modeling will be reviewed. Finally, a more in-depth treatment of statistical modeling choices appropriate for GIS-enabled ecologic studies, some of which are described in the preceding case study sections, will be offered.

Data Quality

The mapped output of a GIS can be extremely persuasive, giving the impression of greater precision than is actually the case. "An increase in the amount of data does not necessarily increase the amount of information. An analysis linking several data layers is only as accurate as the least valid data layer" (Waller, 1996a; Waller, 1996b, p. 86).

Aspects of Accuracy and Precision. Whether using an existing database or designing one for subsequent collection, special attention needs to be paid to the issues of data quality and appropriateness to the application (Briggs and Elliott, 1995). Existing databases can be very useful for investigations in which individual-level data are not available, or for which the cost of collecting data is prohibitive (Frisch et al., 1990). Nonetheless, the quality and utility of available databases from diverse sources vary widely, and databases assembled for their purpose by investigators can be prone to many of the same flaws. Questions which need to be asked of each dataset include (Twigg, 1990; Frisch et al., 1990; Tim, 1995; Vine et al., 1997):

- How current is the information in the database?
- How complete, for the purposes of the application for which it is being considered, is the database?
- What is the lineage of the data; i.e., when were the data collected, what agency collected it, and for what purpose?
- How was the database maintained, how has reality changed out from under it, how will it continue to change, and who will be in charge of database maintenance?
- How accurate are positional and attribute data?
- At which scale was each data layer collected?

As one example, an examination of 26 environmental databases in California revealed numerous drawbacks to their use (Frisch et al., 1990). To be useful in an epidemiologic study, an available database should have accurate locational data (coordinates), temporal data (duration as well as timing of events), and quantitative (exposure measurements) data. Most had one or two of these types of information, but few had all three. Being quite frank in publishing information concerning the accuracy, precision and lineage of all data used can help prevent misunderstandings, although it is no guarantee.

In addition to straightforward questions of accuracy and precision, types of data-related errors arising from unavoidable data manipulation which are likely to affect ecologic studies include measurement error, effect modification (where the meaning of an effect variable changes from one scale to another), and misclassification (where an observation is grouped into the wrong category) (Greenland and Morgenstern, 1989). The common GIS operations of address-matching, interpolation, and "rubbersheeting" (distorting one data layer slightly to make it spatially congruent with another) can all induce measurement error, effect modification and misclassification.

The Hazards of Address-Matching. In databases with individual-level data georeferenced by address, the process of address-matching, if not carefully monitored, can introduce substantial misclassification and measurement error. Matching rural addresses with coordinates usually produces very low initial successful match rates; urban areas with well-maintained databases can often do much better. Vine and colleagues' (1997) initial match rate was 28% in a rural area, which they were able to improve through the use of respondent-marked maps and the most current street database they could find, in this case a set maintained by the school district. Diligent review of match failures in the "Radium City" study permitted a final match rate of 100 per cent, although the first pass produced very poor results (Tobias et al., 1996). Some matching algorithms permit approximate, or probability-based matches, and unless the choices made by such programs are carefully reviewed, the researcher could easily assume the match is much better than it actually is.

Scale Issues. The experience of seeing environment-disease associations appear at one level only to disappear at another (sometimes referred to as the "modifiable areal unit problem" (Openshaw, 1983)) is not uncommon (Openshaw et al., 1987; Knox and Gilman, 1992). GIS can divert trouble if the data are collected at a fine enough resolution. If the goal is to depict the most accurate relationship between putative agent and health effect, the best way to proceed is to collect event (health effect) data and their point locations, and supplement those sets with other attribute data disaggregated as much as possible so that users can define their own small-area boundaries using GIS's redistricting functions. If this course is selected, however, the investigator must take care to select areal units whose scale can faithfully represent the scale at which the pattern or process varies over space. Building areal units from highly disaggregated elements in such a way that within-unit variation is minimized and between-unit variation maximized should produce the highest correlation values possible between model and response (Openshaw, 1978), provided the model is adequately specified (all necessary variables have been identified and are included). Openshaw and colleagues' Geographic Analysis Machine (GAM) permits a boundary-free analysis of data by conducting repeated analyses in an adaptation of the Monte Carlo method, in which moving circular frames of various radii are moved across the study area in an overlapping pattern and observed cases within each frame are counted. This process is repeated as many as 500 or 1,000 times, and each time a frame containing a "significantly higher" number of cases occurs, a circle is drawn on a map. In this way the areas which produce the most highly significant excesses of cases can easily be identified, as well as the range of scales in which statistical sensitivity is greatest (Openshaw et al., 1988).

Sometimes the researcher has no choice but to use predefined administrative boundaries, because available data are pre-aggregated to that level. However, one must critically review the decision to use any boundary system as the unit of analysis in light of and uptake patterns and processes of the substance(s). Unless the areal unit is appropriate to the scale of the patterns and processes under study, administrative boundaries will group higher exposed with lower exposed subjects, introducing ecologic or aggregation bias due to misclassification (Wynder and Stellman, 1992; Greenland, 1992; Carstairs and Lowe, 1986; Cleek, 1979).

Confidentiality

The features that make GIS such a useful instrument also carry substantial ethical risks: its ability to augment health data with spatial information and display it in readily interpretable forms jeopardizes individuals' confidentiality (Vine et al., 1997). U.S. government agencies have expended considerable effort to limit the potential for breach of confidentiality, especially in the health sector. One report cautions, "As geocoded information is added to patient, population and facility-based data files, the value of these files continues to rise sharply, particularly in the private sector" (Croner et al., 1996, p. 1971).

Members of the GIS community have given attention to the issue as well, realizing that only if they voluntarily and effectively protect privacy, "...eventual laws for controlling the detrimental effects of GIS on privacy are less likely to restrict the beneficial uses of GIS or will restrict them to a far lesser extent" (Onsrud et al., 1994, p. 1089). It is recommended that parties within government agencies and in the private sector execute binding agreements "on a case-by-case basis in order to ensure that the public trust is not compromised" (Croner et al., 1996, p. 1971).

Confidentiality can be preserved in a number of ways. In the Radium City individual-level study, final maps are reproduced without street lines (Tobias et al., 1996). The North Carolina Central Cancer Registry has instituted a number of measures designed to prevent parties from identifying cases from published maps. Some of these techniques include offsetting point locations on final maps by an algorithm known only to the agency; others have used this technique as well (Rushton et al., 1995). The N.C. Registry also uses the following methods for published maps: use a scale small enough to conceal identity; map more cases and histotypes than the investigators are interested in; map a several-year collection period, and use the "Rule of Three"—if there are 3 or fewer cases in a cell, aggregate it with a neighboring cell (Aldrich and Krautheim, 1995).

Potential Methodological Pitfalls

Ecologic analysis can actually complicate the task of decomposing disease patterns. Perhaps this is true because it is difficult to think truly ecologically, to compile a sufficiently comprehensive list of potential causative and promotive factors. One often-overlooked complication is the interlocking nature of disease competition, a process that impacts areally-defined populations. Moreover, we are far from announcing that we have the compendium of causes and contributing factors for each histologic type; for these reasons the models of GIS-aided ecologic studies are likely to be dogged by misspecification for some time to come.

Disease competition presents tricky problems of interpretation if a single disease is examined without reference to other diseases with related etiology, "in a vacuum," so to speak (Greenberg, 1985). One classic example is diseases caused by cigarette smoking, in which rises in one smoking-related disease can delete candidates for mortality from the pool of smokers, thereby depressing mortality rates from other smoking-related diseases, which are legion. Smoking is associated with a number of cancer sites, including bladder, oral, nasopharyngeal, stomach, and rectal cancers. If only one disease in the etiologically related system is examined without reference to other competing diseases and to the spatial distribution of promotive factors such as smoking, its mortality trend is likely to be misinterpreted.

Aggregation bias causes an incorrect inference about behavior at the individual level from behavior at another level (Langbein and Lichtman, 1978). It can be caused by misclassification, misspecification, measurement error or effect modification introduced by the aggregating process itself (Greenland and Morgenstern, 1989; Greenland, 1992; Greenland and Robins, 1994; Morgenstern, 1995). This condition acquired the descriptor "ecological fallacy" some time ago, but the name is misleading as it implies that all ecologic studies are to some extent spurious (Susser, 1994a). The so-called fallacy only exists when the inferences drawn are incorrect (Langbein and Lichtman, 1978). The bias works equally well in the other direction: the "atomistic fallacy" reports an incorrect inference about group-level behavior from individual-level data (Susser, 1994a). Causes and cures of aggregation bias are discussed in more detail in the "Modeling Issues" section below.

One important way that aggregation introduces error is through imprecise definition of exposure zones and inadequate identification of the exposed population. For example, assumptions about the nature of diffusion need to be stated explicitly and examined. For instance, when synoptic weather data and atmospheric modeling are unavailable, investigators sometimes use proximity to a hazard as a proxy for exposure (Nuckols et al., 1994; Elliott et al., 1992). Sometimes this is a reasonable substitution, but sometimes it is not (Waller, 1996a). In atmospheric transport situations, for instance, tall stacks injecting emissions high into the atmosphere displace zones of highest deposition some distance from the source (Susser, 1994b; Knox, 1994; Briggs and Elliott, 1995). Likewise, thermally hot emissions from explosions can do the same thing. In the United Kingdom heavy rainfall deposited Chernobyl fallout most heavily in three Welsh counties on the west coast of the island of Britain, farthest away from the accident (Bentham, 1991). Similarly, differences in wind direction, speed and frequency can cause two equidistant receptor zones to have markedly different actual exposures.

Neglecting the influences of pathway, uptake rates and the varying susceptibility of different population subgroups is also likely to introduce substantial measurement error. One case in which population characteristics strongly influence exposure is that of the biologically important radionuclide Iodine-131. Doses of greatest significance of this nuclide are delivered through the milk pathway (Shapiro, 1990). Its eight-day half-life generates very frequent decay events, causing infants to receive by far the highest and most hazardous doses due to their high milk consumption/body weight, small thyroid size, and rapid metabolism and cell division (Shapiro, 1990). For the same reasons of size, metabolic rates and rates of cell division, infants and children can be much more susceptible than adults to the effects of pesticide residues in food and water (Thomas, 1995). Differences due to gender, occupational exposures, advanced age, race, ethnicity or cultural practices such as diet and house type can also generate significant differences in response to a contaminant. Therefore, two areas with exposure to exactly the same concentration of the same substances can have markedly different health impacts due to population structure and cultural factors.

PRINCIPLES OF VALID ECOLOGIC STUDY DESIGN

Accuracy of Ecologic Studies

Ecologic studies can produce less biased estimates than individual-level studies, data problems notwithstanding (English, 1992). For example, one measurement taken of one person's blood pressure can be a poor estimate of his average blood pressure due to daily fluctuations, but single measurements from a number of individuals can provide a very good estimate of the group's mean blood pressure (English, 1992). Nonetheless, aggregation more commonly introduces bias, and usually requires corrective study design for maximum accuracy.

Precise Exposure Modeling

The more precise the model of population exposure, the lower the potential for aggregation bias. It has long been recognized that aggregation bias can be minimized if small areas are grouped in such a way as to minimize within-area and maximize between-area variation of independent or effect variables, as recommended by Openshaw (1983). Richardson and colleagues evaluated the aggregation bias problem and concluded that the greater the difference between group means, the less aggregation bias can influence the findings, provided each mean's variation is not too wide (Richardson et al., 1987).

Surveillance vs. Snapshots

Although change is central to questions of exposure and health impact, few studies performed to date are more than snapshots of current conditions in the environment and the health of its inhabitants. Ongoing data collection permits the development of models and repeated analyses testing different scenarios. GIS technology is an ideal tool to take advantage of such information, as it has the power to manage temporally as well as spatially referenced datasets, and can generate a series of spatial images reflecting the changes in an area's contamination over time. Conceptualizing the environment/public health relationship as one which occurs over temporal space as well as geographic space promotes good database design (Nuckols et al., 1994). Having a longitudinal database handy for assessments of potential hazard also permits more statistical stability for the detection of change.

Thinking Holistically

Cultivating a holistic way of thinking about the ecologies of contamination and disease is probably the best skill an investigator of suspected public health problems can have. It is good to thoughtfully review the list of problem attributes given in the Introduction to this chapter, looking for gaps in one's exposure model. It is also wise to learn as much as possible about the health outcome in question, including what other diseases can be caused by the same suspected contaminant (disease competition), and what other exposures can cause or promote the health outcome under study (such as interaction). If present, data representing these processes need to be included in the exposure model.

HAZARDS

Correctly defining detailed exposure zones and identifying exposed populations have the utmost bearing on every study's ability to detect health impacts. The counts of exactly who is exposed to exactly which hazard(s) in exactly what concentrations and combinations form the terms of the most commonly used epidemiologic measures—relative risks, odds ratios and the standard mortality ratio (Nuckols et al., 1994). Misclassification is likely to seriously impair the study's ability to detect true excesses. To date, few studies have successfully joined all necessary elements to create a truly comprehensive impact assessment. Although many of the following examples of applications of GIS to environmental health lack all of the elements required to assess impact from release to endpoint, they are offered because they execute a few of the required elements well, or tie several elements together in useful ways.

GIS Applications in Risk Evaluation

GIS technology is well suited to the five functions in risk evaluation: scoping, communication, assessment (risk analysis), management, and monitoring (Nyerges et al., 1997). Several examples of the use of GIS in risk assessment, management and monitoring have been given in the Features and Problems sections (above) and the applications described below, but the potential contribution of GIS to the scoping and communication functions is worth including here. The scoping process establishes the extent of concerns relevant to the purpose of the project at hand, and is established by those invited to participate in it. Although some concerned/affected parties (often referred to as stakeholders) have been excluded in scoping processes in the past (Wood and Gray, 1991), this omission can jeopardize the success of the project. New developments in "public participation GIS" are taking advantage of the World Wide Web to disseminate information (Scott and Cutter, 1997).

A GIS application can be adapted to include more people in spatial decision-making, either as consumers of the information, or as actual participants in the decisions. The Spatial Understanding and Decision Support System (SUDSS) concept is evolving as a system that supports spatial decision-making with location-allocation and multi-criteria decision models, based on an Internet-based issue discussion component that enables groups to discuss and evaluate risk-prone situations in both structured and unstructured ways (Jankowski and Stasik, 1996; Moore, 1997). This can prove especially helpful when controversial siting decisions are being made which affect the public's perception of safety and security, such as whether to add a hazardous waste facility or close one down. In questions such as these, GIS implemented within a SUDS System, can be a democratizing and unifying influence in a community.

Mapping Contamination Zones

The U.S. Environmental Protection Agency's (US EPA) Toxic Chemical Release Inventory (TRI) is a rich resource for publicly available pollution data. Releases in the Southeastern United States have been characterized using a GIS at the US EPA (Stockwell et al., 1993). The authors devised a toxicity index profile (TIP) score summarizing the risks associated with each listed chemical, and examined TIP score, frequency and volume according to geographic location and population distribution, concluding that the largest quantities of TRI releases in the Southeast are usually near densely populated areas. Mapping TRI releases provides a visual focus to the relative magnitude of the releases and identifies areas which may need additional study or increased risk management attention (Stockwell et al., 1993). Verification of latitude and longitude coordinates of point sources is necessary, however, as these data are self-reported by industry and have been known to contain errors.

The Agency for Toxic Substances and Disease Registry is using GIS to study groundwater contamination associated with a retired Department of Defense contractor, the former Conductorlab facility in Groton, Massachusetts (Maslia et al., 1994). The chemicals trichloroethylene (TCE), 1,1,1-trichloroethane (TCA), hexavalent chromium, chromium, and lead are the contaminants of concern at this site. The study uses GisPlus software to model its groundwater flow and contaminant transport model. Two transport scenarios were considered in this report, one with continuous toxic input over a 40-year period, and the other with the same condition for the first 20 years, and then a pump-and-treat strategy for the next 20. The resulting concentration surface was then mapped for each scenario. Concentration based on the 40-year scenario was mapped from the temporal perspective: the advance of the 5 ppb maximum contaminant level (MCL)—the "bright line," or boundary between areas contaminated above some regulatory or detection limit, and those below (Graham et al., 1992)—at time slices of 5, 10, 15, 20, 30, and 40 years. The 20-year contamination/20-year remediation scenario was mapped in two lines: the 5 ppb MCL set by the EPA, and a 500-ppb isoline. Through the use of overlay census block data from the Summary Tape Files and TIGER/Line files for block boundaries were used to identify residences at risk. Human exposure values were based on both ingesting water and showering.

Contamination in a body of water can also be mapped using GIS. Twenty-one sampling stations distributed throughout the 27,000 square kilometer area of Lake Erie recorded concentrations of 17 chemicals of concern, including polychlorinated biphenyls (PCBs), dieldrin, lindane and others released by the heavily industrialized communities surrounding the lake. Spatial interpolation was used to calculate concentrations between sampling locations. Overlay analysis and mapping revealed that the East Basin near Buffalo, New York, and fed by the Niagara River, along with the West Basin near Sandusky, Ohio, were the two most heavily contaminated areas of the lake (Wang and Xie, 1994).

Recognizing Environmental Inequity

In 1987 the United Church of Christ released a seminal report on an exploratory study which examined whether communities of color are more exposed to hazardous waste treatment, storage or disposal facilities in general or to commercial hazardous waste landfills, as a special case (United Church of Christ, 1987). Several methods of determining significance of excesses were applied. Geographic scale was quite fine for the time, set at the five-digit zip code level, and five major socioeconomic and demographic variables were examined with regard to presence of a facility of concern within the zip code: minority percentage of the population, mean household income, mean value of owner-occupied homes, number of uncontrolled toxic waste sites per 1,000 persons, and pounds of hazardous waste generated per person. Home values served as a proxy for land values. The "pounds of hazardous waste generated" rate variable was included to accommodate the possibility that the waste site was located near its customers, i.e., factories generating hazardous wastes. A second descriptive study examining the racial and ethnic characteristics of communities in which uncontrolled toxic waste sites were located was also reported.

Although home values were a significant discriminator, minority percentage was more highly correlated with proximity to waste sites. For zip code areas with one operating commercial hazardous waste site, the mean minority percentage was twice that of zips without any such site; for zips with two or more operating facilities, or one of the five largest hazardous waste landfills, the mean minority percentage was more than three times that of the site-free zips. The descriptive study concluded that, although more than half the U.S. population lives in zips containing one or more uncontrolled toxic waste sites, three out of every five Black and Hispanic Americans lived in these communities. Race is the single best predictor of where commercial hazardous waste facilities are located—even when other socioeconomic characteristics such as average household income and average value of homes are taken into account (United Church of Christ, 1987; Mohai and Bryant, 1992).

This report proved to be a shot across the bow of both national and local political and economic forces. The common assumption—that preferential hazardous facility siting was not deliberate but merely an artifact of lower land values—was refuted by the power of the percent minority variable in comparison to the variables of home values and mean household income. Numerous governmental and academic research reports have continued to document the presence of inequities, until in 1994 President Clinton issued an executive order requiring every federal agency to reduce and prevent environmental inequities, and mandating that agencies collect and analyze both pollution and demographic data to determine if their policies are unfair to certain socioeconomic groups or regions. All federal agencies are now required to develop strategies for distributing the pollution burdens equitably and to insure that their policies do not affect one population group or region unevenly (Cutter, 1994). As long as the environment is protected by the federal government, the U.S Constitutional guarantee of equal protection under the law includes environmental protection.

Nonetheless, scientific findings testing the presence of environmental inequities have resulted in widely different conclusions, in spite of the fact that a number of good, rigorous studies have added both evidence and benchmark methodologies to the conclusion that significant inequalities exist (McMaster et al., 1997). GIS has a unique role to play in identifying and characterizing injustice by providing the means to define potentially exposed populations, which simultaneously avoids aggregation bias and improves modeling precision (above; also in Nuckols et al., 1994; McMaster et al., 1997).

Scale and resolution play an especially important part in environmental equity studies (Cutter et al., 1996; McMaster et al., 1997). The United Church of Christ report produced reliable findings with five-digit zip codes as the areal unit of resolution. County-level studies, on the other hand, can be too coarse to reveal patterns of inequitable exposure even when it is there (Cutter, 1994). A study conducted at census tract resolution and using ARC/INFO and S-Plus software to examine the frequency of TRI facilities reports that race is a consistently strong predictor of siting when per capita income and population density are controlled for (Burke, 1993).

Applying the most appropriate methods strengthens the confidence with which results can be regarded, and often strengthens the results as well. The issues of data and measurement quality, scale and resolution, and method of analysis are addressed in a review of GIS equity studies, and accommodated in the equity analysis of minority and poor populations' exposure to TRI emissions in Minneapolis-St. Paul, Minnesota (McMaster et al., 1997). Risk assessment and equity analysis must proceed at multiple scales including the neighborhood to identify locations of disproportionate burden of risk. Type and quantity of hazardous materials are often overlooked, but are an important factor in health impact and should be weighted appropriately. Measures of concentration of poverty, which include information about both poverty and race, can be more effective independent variables than household income or percent minority. In this study, progressively more rigorous renditions of scale, resolution, hazard and disadvantage produced increasingly more robust positive findings (McMaster et al., 1997).

At this point in the development of environmental impact models, findings based on simple proximity analyses (concentric zone models) should be regarded with reservation, while more precise exposure modeling should be employed wherever possible. In one among several examples of the superior accuracy of exposure modeling, a GIS-assisted comparison of circular buffer zones and buffers delineated by a plume footprint reported that findings using the geographic plume analysis produced larger proportions of nonwhites and individuals below the poverty line within the zone of influence than those based on the concentric zone model. A chemical dispersion model using averaged weather data with TRI emissions data, including a measure of emission-specific toxicity and quantity released, generated the plume footprint (Chakraborty and Armstrong, 1997).

In an era of economic dependence on high-risk industries, environmental protection is regarded by many to be an unnecessary expense or a threat to continued economic expansion. However, maintaining a suitable quality of life as well as a competitive edge as an area of potential industrial development requires a public policy of risk reduction. "A concerted effort on the part of the research, regulatory, industrial and public communities to reduce the hazards of this toxic landscape and develop more equitable solutions to these complex problems will go a long way in moving the confrontational politics of jobs versus environment to more longer-term discussions of economic development with environmental justice for all" (Cutter 1994, p. 7).

Siting Facilities With Health Effects Modeling

There are many advantages of using a GIS to site facilities which entail some level of hazard. Factors that GIS technicians involved in these applications often cite include: the complexity of handling many layers of spatial data, the need to accommodate numerous conflicting demands from a variety of stakeholders, the desire to locate the site near needed resources such as transportation routes, water supplies, potential customers and so forth, and the political demands of keeping the selection process as open as possible. Each of these complications can be better managed in the context of a GIS. The high quality cartographic output of GIS also improves communication with the public and political representatives.

Denton County, Texas, has served as the setting for a case study of landfill siting. Old landfills have been filling up, and new landfills have been increasingly difficult to site, as both community size and quantities of municipal garbage generated have grown, environmental regulations have proliferated, and residents have become increasingly sensitive to dump truck traffic, noise, noxious odors, vermin and threats to property values. Environmental and other land use variables were mapped and digitized into a GIS, and each variable was assigned a weight depending on its relative importance as a hazard or other undesirable feature. Spatial analyses were then conducted to identify sites with the least impact and the greatest likelihood of public acceptance, and seven candidate sites were finally chosen to present to the communities and the county government for final selection (Atkinson et al., 1995).

A similar effort mounted in Taiwan emphasized the aspect of communicating with the public to an even greater extent. In addition to making siting-related information available to the general public, this team of engineers used the project to help local environmental protection agencies maintain their own GIS and erected a multimedia World Wide Web interface accessible to anyone (Kao et al., 1997).

Nor are GIS-assisted siting procedures limited to stationary hazards. The Department of Civil Engineering at Iowa State University recently demonstrated a GIS for planning the transportation routes of highly radioactive waste material. Demographic data, environmental features, and transportation system characteristics were included among the required spatial data layers, and three risk-assessment scenarios were run: comparative study, worst-case assessment, and probabilistic risk assessment. The GIS is being used to generate estimates of resident and visitor populations and ecologically sensitive areas along transportation corridors (Souleyrette and Sathisan, 1994).

CONNECTING ENVIRONMENT AND DISEASE

Some studies have used GIS to explore the spatial relationship between exposure to a toxin or carcinogen and a health event believed to be an outcome of that exposure. GIS really come into their own when they are used to link the putative disease agent with health outcome by incorporating environmental data including detailed transport, fate and exposure models with health monitoring evidence such as blood, urine or hair sample assays. This section will cover some of the exposures which concern regulators and the public: pesticides, lead, toxic wastes, and several types of radiation, including electromagnetic (nonionizing) radiation, and four types of ionizing radiation exposure—natural background (including radon), nuclear power generation, nuclear fuel reprocessing, and nuclear weapons production. Although not all studies discussed below employ GIS, each illustrates detailed environmental modeling, health monitoring or both at a noteworthy level of precision. These methods are well suited to GIS application and can contribute strength in that setting.

Pesticides

Currently, the literature presents suggestive, if not conclusive, evidence regarding the relationship of pesticide exposure to health outcome (Pearce, 1989; Wigle et al., 1990; Viel and Richardson, 1991, 1993). The potential for overexposure to herbicides can be reliably identified by a GIS. Hornsby (1992) describes a GIS design which links an environmental fate model to spatial soil and weather data to estimate potential for leaching and runoff. This information is then combined with an extensive database on toxicity and transport parameters for more than 50 herbicides to evaluate risk in farming areas of Florida. Toxicity is represented by the EPA's Lifetime Health Advisory Equivalent (HAL or HALEQ) rating, a "bright line" technique. Thus, the study does not depend either on bioassay results or on environmental monitoring data, which information had already been included in the value of the HAL/HALEQ score.

Lead

The missing link in many exposure/spatial and exposure/GIS studies is evidence of somatic (bodily) damage suspected to have resulted from the exposure in question. To assemble evidence of causality, an approximate dose-response curve should be demonstrable between the exposure map and the somatic damage map, as well as the exposure-to-health outcome map and the somatic damage-to-health outcome map. Somatic damage here would be some kind of physical evidence directly linked to the exposure, such as lead in blood, strontium-90 in bone, aluminum in the brain, plutonium in the lung, and so forth. Health outcome, on the other hand, would be the disease resulting from that specific form of somatic damage. Studies of lead exposure have probably come closest to bringing all three components together, which yields four advantages (Guthe et al., 1992):

- more definitive release – to contamination – to internal deposition – to health outcome models can be derived;
- proxy variables from existing databases which can function as risk markers can be identified;
- running the models with proxies in areas which have not been surveyed for contamination can identify at-risk neighborhoods; and
- situations in which outcome data such as positive blood lead assays or findings of neurological impairment greatly exceed or fall short of predictions can be more closely examined to calibrate the model.

Wartenberg developed a hypothetical example to illustrate the design of a lead screening program which would use GIS to improve the identifying power of screening efforts, and in turn, validate and fine-tune the model (Wartenberg, 1992). Guthe and colleagues (1992) described the use of New Jersey's GIS lead exposure model on greater Newark with environmental sampling, risk markers, and blood screening to demonstrate that the model is effective in identifying at-risk neighborhoods, and that residuals (observed vs. expected) can be mapped to reveal areas where the model works poorly, in order to improve it.

Toxic Wastes

Proximity to hazardous waste sites has long been suspected to be a risk for congenital malformations. One proximity study examined the addresses of newborns and selected 9,313 newborns with malformations and 17,802 matched controls within a one-mile radius of the site, to assess health impact with relation to distance and direction from the site and toxicity of exposure. A slightly higher odds ratio for some defects was detected among cases (Geschwind et al., 1992). Although this study did not have the advantages of GIS technology and was limited to basic proximity analysis, it established the rationale for examining this question further with more powerful technology and more sophisticated modeling. The precision of the one-mile radius limit and the use of distance as a continuous variable probably compensated for the shortcomings of a concentric model to some extent.

Mapping individual birthweights by address is a good demonstration of the enhanced analytic power of examining data in their spatial as well as their numerical domain (Stallones et al., 1992). Mapping birthweight as a continuous variable displays more information than, for instance, choropleth mapping of subareas using low-weight births (<2500 g) as a percent of total births, or dot-mapping addresses of low birthweight events as a dichotomous variable. The authors point out that birthweight is reduced in response to a wide variety of environmental insults, and can thus indicate the overall impact of the locally available chemical soup, rather than relying on individual chemicals and their interactions, which requires more statistical power than many toxically impacted communities can muster. In this study, neither environmental monitoring data nor transport modeling were used, but the method it produced is a useful diagnostic tool for surveillance to select areas for more detailed (and expensive) analysis.

Radiation

Electromagnetic Field (EMF) Radiation. Believing that previous studies had set too low an exposure cutoff, Wartenberg and colleagues (1993) have used GIS to redefine the width of EMF-related health impact near 230 kV ("high tension") electrical transmission lines and to identify proxy variables which can be used as risk markers for model refinement and at-risk neighborhood identification. The investigators did more detailed environmental monitoring than earlier studies to determine a comparatively narrow buffer zone for high-tension lines which would isolate a population receiving an average of about four times the exposure threshold of the earlier work. Because of the exponential decay of EM fields, distance measures from the line are quite adequate as good estimates of exposure, minimizing the need for monitoring fieldwork. Although no bioassay or health outcome data are included, the authors established that within a town, buffer blocks tend to have lower average housing values than the town as a whole. Other variables reflecting perceived housing value, such as percent owner-occupied and average rent, also have the same usefulness as risk markers. The authors ably demonstrate the value of GIS in combination with census variables; studies capable of more carefully delineated exposure strata are possible, which then contribute toward developing a more accurate exposure metric for use in epidemiologic investigations of excess cancer.

Ionizing Radiation Exposure. Ionizing radiation originates from two sources, naturally-occurring background and man-made activities. There is nothing inherently safer about natural background radiation: currently, about 47,600 cancer deaths per year result from exposure to cosmic or terrestrial radiation sources, of which about 32,300 are attributable to inhaled radon and its progeny (National Research Council, 1990; Schleien, 1992).

Background Beta-Gamma Radiation. The Oxford Survey of Childhood Cancers, a case-control study of all children dying from cancer in the United Kingdom between 1953 and 1979 and born between 1944 and 1979, is an invaluable resource for examining questions of environmental carcinogenicity related to children. When the National Radiological Protection Board completed a gridded survey of terrestrial gamma radiation with measurements taken no more than 10 km apart, it became possible to overlay this survey with the Oxford database, which contains address data as well as potentially confounding sociodemographic and medical history data. A significant excess attributable to exposure was detected, along with evidence of interaction effects between magnitude of background exposure and history of prenatal X-rays (Knox et al., 1988). A similar congruence of childhood cancers with background gamma has been detected in the Three Mile Island area (Hatch and Susser, 1990b).

Radon. One example of an ecologic study establishing a dose-response effect between radon exposure and lung cancer examines residents of Iowa, a state with areas of high radon exposure. This hypothesis-generating study examining the interaction of residential radon exposure, smoking and urban/rural status in 20 Iowa counties used mailed surveys to ascertain smoking status, and EPA canister data for radon measurements (Neuberger et al., 1994). The respondent households were stratified into low, medium and high radon strata, and low, medium and high smoking prevalence strata. This method aggregates radon and smoking exposure to the county level, and nine combinations of radon/smoking intensity are obtained. In the low-smoking counties, high-radon areas are associated with much lower lung cancer and small-cell carcinoma incidence rates than low-radon areas, a finding for which the authors had no explanation, but in the high-smoking counties, rates for all lung cancer, adenocarcinoma and small cell carcinoma are significantly higher in the high-radon counties as compared to the low-radon group. Urban counties also report significantly elevated lung cancer rates, after controlling for smoking prevalence. Although radon in the absence of smoking has a relatively small effect, the synergism between radon and smoking has been associated with a high risk for lung cancer in numerous individual-level studies. This study also produced evidence of a radon/urban residence interaction effect, controlling for smoking. Both interactions were evaluated as highly significant in a multivariate analysis technique. This investigation did not employ GIS, but the following review demonstrates a GIS application to this problem.

In an effort to monitor residential exposure to radon in the State of Washington, the State Department of Health is assembling an extensive GIS comprised of data on geology, geography, topography, soil permeability, indoor test results, population density and distribution, and housing. The eastern half of Washington State has some rock formations emitting significant concentrations of radon and its decay products, which are potential public health hazards if new commercial or residential buildings are improperly designed. The health department staff are using the USGS National Uranium Resource Evaluation (NURE) survey data for detailed beta-gamma measurements made on a fairly fine scale (90 meters apart). Their goal is to satisfy local health districts' need for a cohesive public policy on radon despite the persistent controversy about its health effects and doubts about federal radon policies (Coleman et al., 1994). When linked to disease and smoking data, this GIS will provide an effective ongoing surveillance tool.

Nuclear Power Generation. No genuinely GIS-enabled examination of health effects resulting from routine operation of nuclear power plants has been conducted outside the context of an accident. The health of communities surrounding the Three Mile Island nuclear plant after its partial core meltdown in March of 1979 has been examined in several studies, although in most cases inadequate time has elapsed to expect to see health effects from that accident (Hatch et al., 1990a). In these studies, areal units were assembled into study blocks from census blocks within a 10-mile radius of the plant. Covering incident cancers between 1975 and 1985, the investigators found slight increases in childhood leukemia and childhood cancers, but the numbers were low and the estimates highly variable. Other rises were noted in non-Hodgkin's lymphoma relative to both accident and routine emissions, and in lung cancer relative to accident emissions, routine emissions, and background gamma radiation. Given the brief duration of the study period, it is not surprising that this study failed to produce convincing evidence that radiation releases from that accident "influenced cancer risk during the limited period of follow-up" (Hatch et al., 1990a). These results are no guarantee that significant excesses resulting from TMI operations will not emerge once the established latency periods have elapsed.

Nuclear Fuel Reprocessing. Childhood cancers and leukemias have been repeatedly examined with a variety of techniques near the Sellafield (formerly Windscale) spent fuel reprocessing station in West Cumbria, United Kingdom. In the boundary-free technique mentioned in "Scale Issues" above, Openshaw and colleagues (1987; 1988) adapted a GIS containing the cancer registries of Newcastle and Manchester with the novel Geographical Analysis Machine (GAM). As described above, this device drew lagged circles over the study area, counting the number of cancers falling within the circle each time. Circles demonstrating excess cases occurred 1792 times, where a Poisson distribution would have expected 173. The two main areas of excess cases were located in Seascale and Tyneside. The advantages of the GAM included that the machine could be set to run and count with no prior hypotheses established, and simply detect and report the excesses it located. Nonetheless, the investigators were operating in the absence of environmental monitoring data, weather data, diffusion modeling, and bioassay data, with the principal proposed risk marker being proximity.

Nuclear Weapons Production. The most advanced exposure/GIS projects being developed at the time of this writing are the Dose Reconstruction Projects for populations surrounding the nuclear weapons production facilities of the Hanford Reservation (Washington State) and the Savannah River Site (South Carolina). The Hanford project, being conducted jointly by Battelle Pacific Northwest Laboratories and Risk Assessments Corporation (RAC, formerly Radiological Assessments Corporation). RAC is also involved in the Savannah River Site project. The dose reconstruction projects lack bioassay data from the nonworker public, but they surpass the lead model in that they account for all pathways imaginable, from eating fish through drinking milk from a family cow pastured near the plant boundaries, to incorporating some of the most advanced atmospheric dispersion modeling available to date. The Hanford study is nearest completion, and contains three main components: the Columbia River pathway (Farris et al., 1994b), the Atmospheric Transport pathway (Farris et al., 1994a), and the Regional Atmospheric Transport Code (RATCHET) (Ramsdell et al., 1994). Atmospheric Transport takes inputs from RATCHET and conducts them through surface-related pathways such as milk, locally grown vegetables, surface streams, immersion, skin absorption, and so forth.

Although environmental monitoring at the time Hanford and Savannah River were producing nuclear weapons was less than adequate for today's requirements, this deficiency has been made up for as much as possible by reevaluating all plant operations from start-up and recalculating emissions ("source terms"). The initial code has been custom-written by the project, but coverages generated have been integrated on a PC ARC/INFO platform. Initial results have established that early operations of the Hanford plant did indeed deliver extremely high radioiodine doses to infants. The studies are expected to continue for some years, and findings related to exposure to airborne noble gases and transuranic particles are anticipated.

MODELING ISSUES

The Ecologic Fallacy Revisited

"The naming of names," Susser has observed, "often influences attitudes and thought." Selvin's styling of problems related to cross-level (multi-scale) inferences as the "ecological fallacy" (Selvin, 1958) has "brought the ecological approach into disrepute" (Susser, 1994a, p. 825). The potential for becoming entangled in the so-called ecologic fallacy looms near practically every study design suggested here, and those designing ecologic studies are well advised to become familiar with the issues of scale which impact on the inferences they would like to test.

Briefly described, "ecologic fallacy" is a catch-all term for errors in results obtained by making inferences from data collected at one scale to individuals or communities aggregated at other scales. When the scale of analysis shifts, it is not uncommon for the meaning of a variable to change as well. Consider the example of the "hung jury": its answer is inconclusive, but the individual jury members are thoroughly convinced on both sides of the issue, so convinced in fact that they cannot be persuaded to change their minds (Zito, 1975). An ecologic fallacy would be committed if one were to attribute the jury's indecisiveness to each of its members.

It is also common for additional confounding variables to come into play as the scale shifts, causing a perfectly specified individual-level model to become misspecified. For instance, consider two individuals who have identical incomes. The impact of poverty on the one who lives in a deprived neighborhood with poor social networks differs from that on the one who lives in an orderly neighborhood with good social networks. At the aggregate level, the strength of the actual damage poverty inflicts on its survivors will vary widely between deprived neighborhoods and orderly ones, although their census-block economic variables may have equal values. Nonetheless, the neighborhood-level impact is not going to be a perfect predictor of the individual-level impact, as ameliorating conditions will vary greatly from one family to another, and even from one individual to another. Careful attention to the constructs being measured at each level will aid investigators in correctly specifying models (Schwartz, 1994), but there are also choices of method which can to some extent minimize the impact of such errors.

Some authors prefer the term "aggregation bias" to ecologic bias because there is nothing inherently biased about ecologic studies. As in the examples above, from the group perspective, "cross-level bias stems either from the 'atomistic fallacy' inherent in individual observations that ignore group effects, or from specification bias to which individual analyses are also prone" (Susser, 1994a, p. 829).

Misspecification, misclassification, effect modification and measurement error, as described above, are the likely culprits when aggregation introduces bias into an ecologic study. Sometimes aggregation produces parameter estimates of individual-level behavior that are less biased than estimates from individual level data (the blood pressure example above); this is "aggregation gain" (Langbein and Lichtman, 1978). More often, however, aggregation results in a loss of real explanation that may not be evident from the results of the statistical procedures. There are two ways individuals may be aggregated which have potential for aggregation bias (Langbein and Lichtman, 1978):

- Grouping according to values of the dependent (outcome) variable y;
- Grouping according to values of a variable correlated with both the independent (effect) variable x and with y.

Although these grouping schemes can introduce aggregation bias, a correlation analysis should identify the overly correlated variables. If run before the final analysis model is set, the grouping schemes may be rearrangeable, so the data can be more effectively partitioned.

Misspecification. A model is said to be properly specified when the error term, representing unexplained variance, is uncorrelated with any of the independent variables. No variable is excluded which is related to either the dependent or to another independent (Langbein and Lichtman, 1978). Clearly, a model may be misspecified whether or not the data have been grouped. An additional correlation analysis between the independent variables and the residuals of the trial models should uncover faulty specification of this type.

Misspecification does not always have to entail positive covariation. Omitting a competing disease whose covariation with the response variable is influential but negative, is also an example of misspecification. Given the multifactorial nature of cancer causation, avoiding misspecification perfectly is probably not possible given the current knowledge base. When grouping causes to become important at the aggregate level variables which were not present at the individual level, or changes the meaning of variables, the grouping process itself has caused misspecification. As have Openshaw (1978) and Richardson and colleagues

(1987) in other disciplines (above), Greenland and Morgenstern (1989, p. 273) observe that "ecological bias due to model misspecification is inversely proportional to the between-group variation in disease rates." Thus, if the original areal units of analysis are disaggregated enough, the grouping process can result in more between-unit than within-unit variation in outcome variable. Of course, when this kind of grouping succeeds, it does so because the levels of the independent(s) have partitioned the study population based on exposure to risks strong enough to drive the disease distribution.

Misclassification. When effect and response variables are dichotomous or categorical, misclassification can produce disproportionately large errors. Inaccurate disease diagnoses or measures of exposure to risk can readily produce these results. In such cases, an undercount of true positives leads to an inflation of false negatives (and vice versa), so one error can strike twice. Another means by which ecologic estimates are made especially sensitive to misclassification is in the matter of time factors such as changes in exposure rates over time, latency periods, and migration (Greenland, 1992). Misclassification of exposure and disease have been identified as two of the principal errors in epidemiologic studies concerning hazardous waste sites (National Research Council, 1991).

Misclassification related to latency periods is an especially troublesome aspect of data collection and verification. Most diseases resulting from exposures of concern to regulators have latency periods on the order of seven to 40 years. Migration of exposed persons out of the area usually results in these persons being lost to the study. An alternative to dealing with long latency periods is to identify a biomarker or early health effect associated with the exposure. In addition to minimizing this source of misclassification bias, selecting such an early exposure indicator also allows time for intervening action to protect the public health (Nuckols et al., 1994).

Effect Modification, a source of the potential bias in ecologic studies discussed above, occurs when the grouping method alters the effect of an independent variable when moving from the individual to the macro level (Greenland and Morgenstern, 1989). The alteration can occur because the exposure effect varies across groups, due to a differential distribution of extraneous risk factors across groups. The factor need not be a confounder at the individual level. Grouping into counties means that counties will vary widely in, for example, racial composition or socioeconomic characteristics. Race and socioeconomic variables are well-established confounders for many diseases, including cancers. Retaining these variables in the model may be only a partial solution: "[E]cological control of confounders and other covariates responsible for ecological bias cannot be expected to completely remove the biases such covariates produce, and may even worsen bias" (Greenland and Morgenstern, 1989, p. 269). Using study designs, such as multilevel modeling, can maintain the distinction between the individual level effect and the aggregated effect, representing disease patterns more accurately when effect modification is a problem.

Measurement Error. One of the assumptions of the most commonly used statistical techniques, including linear regression (measuring the relation between the value of an outcome variable and corresponding values of effect variables) is that the variables are measured without error. In nature, this assumption is rarely met.

As just one example, the completeness of the underlying cause of death entry on death certificates has varied widely over time, across space, and even from one coroner or medical examiner to another. Racial bias has been one of the strongest factors impacting death certificate accuracy and completion in the United States (Mannino et al., 1996; Coultas and Hughes, 1996; Mannino et al., 1997). U.S. states have improved compliance with federal standards at different rates of progress. In the case of U.S. racial bias, undercount has been much more common than overcount, biasing coefficients toward the null. The problem can be especially acute, but undetected, in studies of rare diseases relying on data prior to the 1980s. To a degree, some methods such as structural equation modeling (described below) can accommodate measurement error and should be used, if feasible.

Appropriate Methods for Cross-Level Analyses

Ecologic study design frequently includes elements of between-zone comparisons and change over time. Ordinary least squares regression assumes that each observation is independent of its neighbors, or in other words, that the magnitude of an effect on one observation does not influence its effect on another. When diseases are spread by contagion or caused by an environmentally transported pollutant, then clearly the intensity of the disease will vary over space, with more exposed persons clustered together. When the trend of a health effect is observed in the same area over time, the magnitude of the effect in one year (or day, or month) can influence the effect's magnitude in subsequent years, either by reducing the size of the susceptible population by removing potential cases (and turning them into actual cases or immune individuals), or by reflecting the presence of an environmental factor to which the surviving population will continue to be susceptible. Autocorrelation reduces overall variation among the observations, which biases the results, especially measures of significance.

The following model designs address these shortcomings in various ways. Although all of these models can be represented in equation notation, statisticians have developed two alternative methods, matrix notation and path diagrams. Path diagrams in particular make the relations within the model easy to visualize and manipulate. In path diagrams, rectangles represent observed variables and circles, unobserved or latent variables. Arrows indicate the movement of effect, usually from independent to dependent. A two-headed arrow reflects correlation. Figure 5.2 depicts an ordinary least squares regression model with one independent variable and no violations of independence. Figures 5.3 and 5.4 are path diagrams of simple mixed and structural equation models discussed here.

Mixed models. Ordinary least squares models apply one intercept to a body of data. If the data are divided into subareas such as exposure zones, a different slope may be generated for each zone, but the same intercept will originate the trends for all the zones. In this case, the effect variables are considered to be "fixed effects." When a model is specified with random effects as well, it becomes "mixed." Mixed models, sometimes referred to as hierarchical linear models or nested designs, are generalizations of the general linear model in which the zones can be named as random effects, and the model calculates a separate intercept, mean

Figure 5.2. Ordinary Least Squares Regression with One Independent Observed Variable. Observed Dependent Variable = Observed Independent Variable + Unobserved Error.

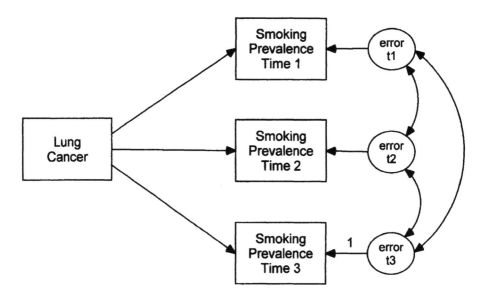

Figure 5.3. Mixed Model Controlling for Temporal Autocorrelation. Observed Dependent Variable = Observed Independent Variable + Unobserved Errors. The independent variable has a dimensional subscript of 1 to 3. The curved two-headed arrows represent correlation. Graphics software: Arbuckle, 1997.

(and slope(s) when appropriate) for each zone (Laird and Ware, 1982; Louis, 1988; Littell et al., 1996). The ability for each zone to be freed of other zones, and to have its own intercept and coefficients calculated, is the "random effects" part of the mixed model. In the case of exposure zones here, specifying ZONE as a random effect controls for the spatial autocorrelation within each zone, and allows between-zone variation to be separated (or "partialled out") from the within-zone variation (Littell et al., 1996).

Where trend over time is the topic of the hypothesis tests, specifying as a random effect the areal unit of measurement for each time point controls for temporal autocorrelation (Laird and Ware, 1982; Cnaan et al., 1997). Figure 5.3 illustrates a simple three time-point model of lung cancer in one county with temporal autocorrelation controlled for.

For a more complex example, consider that the previous example of exposure zones contains several counties in each zone, and that several years of measures were taken on each county. In that case, both temporal and spatial autocorrelation can be controlled for by specifying both zone and county as random effects (Louis,

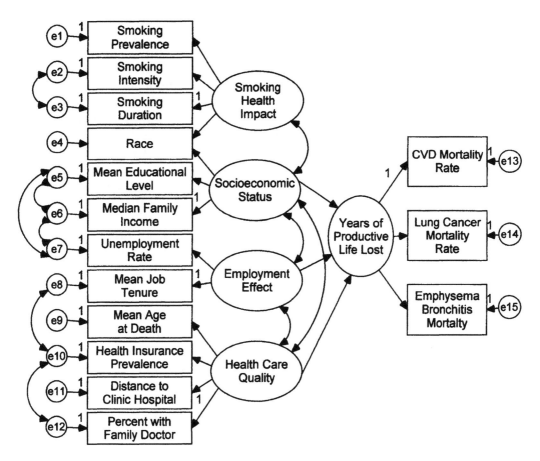

Figure 5.4. Structural Equation Model of Years of Productive Life Lost. Observed Indicator Dependent Variables with Errors = Unobserved Dependent Variable ≈ Four Unobserved Independent Latent Factors measured by (≈) 12 Observed Independent Variables with Correlated Measurement Errors. Graphics software: Arbuckle, 1997.

1988). (If YEAR were specified, one would get a separate intercept and set of slopes for every year in the study, which can produce a blizzard of information which is usually unnecessary.) Mixed models do not correct for measurement error or some of the other sources of aggregation bias discussed above, but in many cases they can be the best choice for the types of questions which need to be answered, including models with interaction terms and more than a few time points.

Multi-level models. Multi-level models are a generalization of the mixed-effects model tailored for cross-level analyses, in which two or more levels can be specified, and the coefficient of any effect variable can be treated as a random variable at any level in the hierarchy (Goldstein, 1999). In some ecologic studies, several levels of aggregation are available and can be used to partition the variation in the data most accurately. For instance, suppose one has both individual-level and aggregate data on an environmentally-related disease for a large area in which there are communities within counties, and counties within regions. Let us further specify that there are cultural and ethnic differences from one region to the next, having to do with which ethnic group settled the region originally. Each community is somewhat autonomous in the ways its clinics and health department choose to provide health care and environmental surveillance.

Whether an individual gets diagnosed, how early he or she is diagnosed, and how much care he/she receives are all contingent on the clinic by which the person is served, and the commitment of his/her community's health department to surveillance. The effects these community-level differences have on severity and outcome of the disease at the individual level can be considered the "ward" or "neighborhood" effect of that community (Humphreys and Carr-Hill, 1991). Several communities are contained in each county, which introduces another level of effect. For the sake of illustration, let us say that there is considerable inter-county difference in socioeconomic status, with some counties enjoying the benefits of higher educational levels and incomes than others. Because it is specified as a random variable with its own probability distribution, the ward effect at the county level captures such inter-county differences. The ethnically distinct multi-county regions are the highest level in this design, with health-related cultural characteristics such as diet, house type, and genetic susceptibility and resistance contributing a ward effect at the regional level. Analyses can be performed independently at each resolution from the individual to the region, but multi-level analyses can also be conducted at higher levels (scales) with the lower levels of data collection (individual, clinic, community, county) serving as units of resolution. One could, for instance, produce regional-scale findings using data collected at individual resolution, and employing intermediate levels variables collected at community and county levels as components of the ward effect.

When multiple levels are present in data but the study design does not accommodate them, a great deal of information is lost, and the differences in poverty effects on health outcome at each level are dropped to the lowest unit of resolution. As a result, some of the variation attributed to each individual-level observation truly belongs to the clinic, community, county or region. In addition to controlling for spatial and temporal autocorrelation, multi-level models also accommodate effect modification by partialling effect variation to each of the scales in which it belongs, if the effect data have been collected at those scales.

Structural equation modeling. First achieving broad application in the field of sociology, structural equation modeling is especially suited to situations in which correlated errors, unobserved exogenous factors, recursive relationships among variables, measurement error and interlocking causality can defeat ordinary multiple regression. Autocorrelation is an example of correlated errors. Figure 5.4 illustrates a structural equation model of years of productive life lost (YPLL) as the outcome (dependent) variable, with unobserved exogenous (independent) factors and recursive inter-factor relationships.

YPLL is an unobserved endogenous factor, with three observed endogenous indicator variables (the mortality rates). On the independent side, unobserved factors are latent variables composed of several observed variables. They are referred to as "exogenous" when they are expressions of data on the independent or effect variable side of the model.

Environmentally associated diseases, including cancers, are rife with these complications. Though fictional, the following example (adapted from Greenland and Morgenstern, 1989) is typical of recursive relationships and interlocking causality. Suppose an unknown dietary factor (perhaps vitamins A or C) reduces smokers' susceptibility to lung cancer. Suppose further that there is county-to-county variation in the consumption of foods and supplements containing these vitamins (a likelihood). Suppose the unknown dietary factor has a differential effect related to a race-linked genetic component. Add to that the realities that there are racial differences in the smoking habit and in the magnitude of health effects/dose. Whites are usually documented as preferring more cigarettes with less tar, while Blacks smoke fewer cigarettes but choose higher-tar brands. Evidence is also accumulating that Americans of African descent are more vulnerable to smoking's health effects than whites, holding doses equal (Caraballo et al., 1998). Finally, educational level further confounds the power of the effect, as documented in Wagenknecht et al. (1990). Parts of the variation belonging more properly to brand preference, number of cigarettes smoked, genetic makeup and diet composition at the individual level would appear as parts of race, county, urban status, household family income, educational level, and so forth. Further, these variables would display considerable covariation and probably some collinearity as well. This kind of multiple, interlocking causality is common in cancer induction. Covariance structure techniques such as path analysis and structural equation modeling specifically address recursive relationships among the data and simultaneous causality (Long, 1983). Campbell and colleagues (1986) recommend structural equation modeling to cope with measurement and misspecification errors, and point out the technique is particularly well-suited for examining processes in which effect variables display covariation and can be mediated by intervening variables.

CONCLUSION

Even when aided with GIS techniques, environment-disease investigations present numerous obstacles:

- demanding methods,
- voluminous data requirements,
- imperfect measurement,
- incomplete understanding of risk factors,
- scale combinations that are less than compatible, and
- that pesky grey area between association and causation, to name a few.

However, we will never get to the point where we can confidently specify a robust model from a comprehensive knowledge base unless we pursue the scraps of information and experience the mistakes of conducting cancer prevention and control with the tools at hand.

While the epidemiologists' maxim, "Where there is doubt, err on the side of saving lives" may oversimplify the kinds of problems health surveillance professionals face when examining problems potentially related to environmental contamination, the viewpoint this maxim supports is still helpful. Sir Austin Bradford Hill contributed the Postulates, which today are the standard to be met when deciding causation in disease-environment relations including the case of cancer. But even Hill concluded: "All scientific work is incomplete—whether it be observational or experimental. All scientific work is liable to be upset or modified by advancing knowledge. That does not confer upon us a freedom to ignore the knowledge we already have, or to postpone the action that it appears to demand at a given time" (Hill, 1965, p. 300).

REFERENCES

Aldrich, T.E. and K.R. Krautheim. 1995. Protecting confidentiality in small area studies. *Proceedings of the Symposium on Statistical Methods 1995.* Centers for Disease Control and Prevention; Agency for Toxic Substances and Disease Registry; Emory University School of Public Health, Biostatistics Division; University of Georgia, Department of Statistics; and the American Statistical Association, Atlanta Chapter, sponsors. Atlanta, GA.

Arbuckle, J.L. 1997. *AMOS (Analysis of Moment Structures).* Ver 3.6. Covariance structure model programming and graphing software. Chicago: Smallwaters Corporation.

Atkinson S.F., F.A. Schoolmaster, D.I. Lyons, and J.M. Coffey. 1995. A geographic information systems approach to sanitary landfill siting procedures: A case study. *Environmental Professional* 17(1):20–26.

Bentham, G. 1991. Chernobyl fallout and perinatal mortality in England and Wales. *Social Science and Medicine* 33(4):429–434.

Bollen, K.A. 1989. *Structural Equations with Latent Variables.* New York: John Wiley & Sons.

Briggs, D.J. and P. Elliott. 1995. The use of geographical information systems in studies on environment and health. *World Health Statistics Quarterly* 48(2):85–94.

Burke, L.M. 1993. Race and environmental equity: A geographic analysis in Los Angeles. *Geo Info Systems* 3(9):44, 46–47, 50.

Campbell, R.T., E. Mutran, and R.N. Parker. 1986. Longitudinal design and longitudinal analysis: Comparison of three approaches. *Research on Aging* 8(4):480–502.

Caraballo, R.S., G.A. Giovino, T.F. Pechacek, P.D. Mowery, P.A. Richter, W.J. Strauss, D.J. Sharp, M.P. Eriksen, J.L. Pirkle, and K.R. Maurer. 1998. Racial and ethnic differences in serum cotinine levels of cigarette smokers. Third National Health and Nutrition Examination Survey, 1988–1991. *Journal of the American Medical Association* 280(2):135–139.

Carstairs, V. and M. Lowe. 1986. Small area analysis: Creating an area base for environmental monitoring and epidemiological analysis. *Community Medicine* 8(1):15–28.

Chakraborty, J. and M.P. Armstrong. 1997. Exploring the use of buffer analysis for the identification of impacted areas in environmental equity assessment. *Cartography and Geographic Information Systems* 24(3):145–157.

Cleek, R.K. 1979. Cancers and the environment: The effect of scale. *Social Science and Medicine* 13(D):241–247.

Cnaan, A., N.M. Laird, and P. Slasor. 1997. Using the general linear mixed model to analyse unbalanced repeated measures and longitudinal data. *Statistics in Medicine* 16(20):2349–2380.

Coleman, K.A., G.C. Hughes, and E.J. Scherieble. 1994. Where's the radon? The geographic information system in Washington State. *Radiation Protection Dosimetry* 56(1–4):211–213.

Coultas, D.B. and M.P. Hughes. 1996. Accuracy of mortality data for interstitial lung diseases in New Mexico, USA. *Thorax* 51(7):717–720.

Croner, C.M., J. Sperling, and F.R. Broome. 1996. Geographic information systems (GIS): New perspectives in understanding human health and environmental relationships. *Statistics in Medicine* 15(17–18):1961–1977.

Cutter, S., L. Clark, and D. Holm. 1996. The role of geographic scale in monitoring environmental justice. *Risk Analysis* 16(4):517–526.

Cutter, S.L. 1994. The burdens of toxic risks: Are they fair? *B&E Review* 41(1):3–7.

Elliott, P., J.A. Beresford, D.J. Jolley, S.H. Pattenden, and M. Hills. 1992. Cancer of the larynx and lung near incinerators of waste solvents and oils in Britain. In *Geographical and Environmental Epidemiology. Methods for Small-Area Studies*. P. Elliott, J. Cuzick, D. English, and R. Stern (Eds.), pp. 359–367. New York: Oxford University Press.

English, D. 1992. Geographical epidemiology and ecological studies. In *Geographical and Environmental Epidemiology. Methods for Small-Area Studies*. P. Elliott, J. Cuzick, D. English, and R. Stern (Eds.), pp. 3–13. New York: Oxford University Press.

Farris, W.T., T.A. Ikenberry, B.A. Napier, D.B. Shipler, P.W. Eslinger, and J.C. Simpson. 1994a. *Atmospheric Pathway Dosimetry Report, 1944–1992: Hanford Environmental Dose Reconstruction Project*. Battelle Pacific Northwest Laboratories. PNWD-2228 HEDR. Richland, WA.

Farris, W.T., S.F. Snider, B.A. Napier, D.B. Shipler, and J.C. Simpson. 1994b. *Columbia River Pathway Dosimetry Report, 1944–1992*. Battelle Pacific Northwest Laboratories. PNWD-2227 HEDR. Richland, WA.

Frisch, J.D., G.M. Shaw, and J.A. Harris. 1990. Epidemiologic research using existing databases of environmental measures. *Archives of Environmental Health* 45(5):303–307.

Geschwind, S.A., J.A.J. Stolwijk, M. Bracken, E. Fitzgerald, A. Stark, C. Olsen, and J. Melius. 1992. Risk of congenital malformations associated with proximity to hazardous waste sites. *American Journal of Epidemiology* 135(11):1197–1207.

Glick, B.J. 1979. The spatial autocorrelation of cancer mortality. *Social Science and Medicine* 13(D):123–130.

Goldstein, H. 1999. *Multilevel Statistical Models*. London: E. Arnold Publishers.

Graham, J., K.D. Walker, M. Berry, E.F. Bryan, M.A. Callahan, A. Fan, B. Finley, J. Lynch, T. McKone, H. Ozkaynak, and K. Sexton. 1992. The role of exposure databases in risk assessment. *Archives of Environmental Health* 47(6):408–420.

Greenberg, M. 1985. Cancer atlases: Uses and limitations. *The Environmentalist* 5(3):187–191.

Greenland, S. 1992. Divergent biases in ecologic and individual-level studies. *Statistics in Medicine* 11(9):1209–1223.

Greenland, S. and H. Morgenstern. 1989. Ecological bias, confounding, and effect modification. *International Journal of Epidemiology* 18(1):269–274.

Greenland, S. and J Robins. 1994. Invited commentary: Ecologic studies—biases, misconceptions, and counterexamples. *American Journal of Epidemiology* 139(8):747–760.

Guthe, W.G., R.K. Tucker, E.A. Murphy, R. England, E. Stevenson, and J.C. Luckhardt. 1992. Reassessment of lead exposure in New Jersey using GIS technology. *Environmental Research* 59(2):318–325.

Hatch, M.C., J. Beyea, J.W. Nieves, and M. Susser. 1990a. Cancer near the Three Mile Island nuclear plant: Radiation emissions. *American Journal of Epidemiology* 132(3):397–412.

Hatch, M. and M. Susser. 1990b. Background gamma radiation and childhood cancers within ten miles of a U.S. nuclear plant. *International Journal of Epidemiology* 19(3):546–552.

Hill, A.B. 1965. The environment and disease: Association or causation? *Proceedings of the Royal Society of Medicine* 58(5):295–300.

Hornsby, A.G. 1992. Site-specific pesticide recommendations: The final step in environmental impact prevention. *Weed Technology* 6(3):736–742.

Humphreys, K. and R. Carr-Hill. 1991. Area variations in health outcomes: Artefact or ecology. *International Journal of Epidemiology* 20(1):251–258.

Jankowski, P. and M. Stasik. 1996. Architecture for space and time distributed collaborative spatial decision making. *GIS/LIS '96*, Denver, CO.

Kao, J.-J., H.-Y. Lin, and W.-Y. Chen. 1997. Network geographic information system for landfill siting. *Waste Management and Research* 15(3):239–253.

Knox, E.G. 1994. Leukaemia clusters in childhood: Geographical analysis in Britain. *Journal of Epidemiology & Community Health* 48(4):369–376.

Knox, E.G. and E. Gilman. 1992. Leukaemia clusters in Great Britain. 2. Geographical considerations. *Journal of Epidemiology & Community Health* 46(6):573–576.

Knox, E.G., A.M. Stewart, E.A. Gilman, and G.W. Kneale. 1988. Background radiation and childhood cancers. *Journal of Radiological Protection* 8(1):9–18.

Laird, N.M. and J.H. Ware. 1982. Random-effects models for longitudinal data. *Biometrics* 38:963–974.

Langbein, L.I. and A.J. Lichtman. 1978. *Ecological Inference*. Sage Publications, Beverly Hills, CA.

Littell, R.C., G.A. Milliken, W.W. Stroup, and R.D. Wolfinger. 1996. *SAS System for Mixed Models*. SAS Institute, Inc., Cary, NC.

Long, J.S. 1983. *Covariance Structure Models: An Introduction to LISREL*. Sage Publications, Thousand Oaks, CA.

Louis, T.A. 1988. General methods for analysing repeated measures. *Statistics in Medicine* 7:29–45.

Mannino, D.M., C. Brown, and G.A. Giovino. 1997. Obstructive lung disease deaths in the United States from 1979 through 1993: An analysis using multiple-cause mortality data. *American Journal of Respiratory and Critical Care Medicine* 156(3, pt. 1):814–818.

Mannino, D.M., R.A. Etzel, and R.G. Parrish. 1996. Pulmonary fibrosis deaths in the United States, 1979–1991. An analysis of multiple-cause mortality data. *American Journal of Respiratory Critical Care Medicine* 153:1548–1552.

Maslia, M.L., M.M. Aral, R.C. Williams, A.S. Susten, and J.L. Heitgerd. 1994. Exposure assessment of populations using environmental modeling, demographic analysis, and GIS. *Water Resources Bulletin* 30(6):1025–1041.

McMaster, R.B., H. Leitner, and E. Sheppard. 1997. GIS-based environmental equity and risk assessment: Methodological problems and prospects. *Cartography and Geographic Information Systems* 24(3):172–189.

Mohai, P. and B. Bryant. 1992. Race, poverty and the environment. The disadvantaged face greater risks. *EPA Journal* 18(1):6–8.

Moore, T.J. 1991. Application of GIS technology to air toxics risk assessment: Meeting the demands of the California air toxics "Hot Spots" Act of 1987. *GIS/LIS '91*, Atlanta, GA: American Congress on Surveying and Mapping, American Society for Photogrammetry and Remote Sensing, Association of American Geographers, Urban and Regional Information Systems Association, and AM/FM International.

Moore, T.J. 1997. GIS, society and decisions: A new direction with SUDSS in COM-MAND. *Proceedings of Auto-Carto 13*, Bethesda, MD: American Congress on Surveying and Mapping.

Morgenstern, H. 1995. Ecologic studies in epidemiology: Concepts, principles, and methods. *Annual Review of Public Health* 16:61–81.

National Research Council. 1990. *BEIR V: Health Effects of Exposure to Low Levels of Ionizing Radiation*. Committee on the Biological Effects of Ionizing Radiation. Washington, DC.

National Research Council. 1991. *Environmental Epidemiology: Public Health and Hazardous Wastes*. National Academy Press. Washington, DC.

Neuberger, J.S., C.F. Lynch, B.C. Kross, R.W. Field, and R.F. Woolson. 1994. Residential radon exposure and lung cancer: Evidence of an urban factor in Iowa. *Health Physics* 66(3):263–269.

Nuckols, J.R., J.K. Berry, and L. Stallones. 1994. Defining populations potentially exposed to chemical waste mixtures using computer-aided mapping analysis. In *Toxicology of Chemical Mixtures*, R.S.H. Yang (Ed.), pp. 473–504. Orlando, FL: Academic Press.

Nyerges, T., M. Robkin, and T.J. Moore. 1997. Geographic information systems for risk evaluation: Perspectives on applications to environmental health. *Cartography and Geographic Information Systems* 24(3):123–144.

Onsrud, H.J., J. Johnson, and X. Lopez. 1994. Protecting personal privacy in using geographic information systems. *Photogrammetric Engineering and Remote Sensing* 60(9):1083–1095.

Openshaw, S. 1978. An empirical study of some zone-design criteria. *Environment and Planning A* 10:781–794.

Openshaw, S. 1983. *The Modifiable Areal Unit Problem*. Concepts and Techniques in Modern Geography, vol. 38. Norwich, England: Geo Books.

Openshaw, S., M. Charlton, C. Wymer, and A. Craft. 1987. A Mark 1 geographical analysis machine for the automated analysis of point data sets. *International Journal of Geographical Information Systems* 1(4):335–358.

Openshaw, S., A. Craft, M. Charlton, and J.M. Birch. 1988. Investigation of leukaemia clusters by use of a geographical analysis machine. *Lancet* February 6, 1988:272–273.

Pearce, N.E. 1989. Phenoxy herbicides and non-Hodgkin's lymphoma in New Zealand: Frequency and duration of herbicide use. *British Journal of Industrial Medicine* 46(2):143–144.

Ramsdell, J.V., C.A. Simonen, and K.W. Burk. 1994. *Regional Atmospheric Transport Code for Hanford Emission Tracking (RATCHET)*. PNWD-2224. Richland, WA: Pacific Northwest Laboratory.

Richardson, S., I. Stücker, and D. Hémon. 1987. Comparison of relative risks obtained in ecological and individual studies: Some methodological considerations. *International Journal of Epidemiology* 16(1):111–120.

Rushton, G., D. Krishnamurti, R. Krishnamurthy, and H. Song. 1995. A geographic information analysis of urban infant mortality rates. *Geo Info Systems* 5:52–66.

Schleien, B. (Ed.). 1992. *The Health Physics and Radiological Health Handbook* Silver Spring, MD: Scinta, Inc.

Scholten, H.J. and M.J.C. de Lepper. 1991. The benefits of the application of geographical information systems in public and environmental health. *World Health Statistics Quarterly* 44(3):160–170.

Schwartz, S. 1994. The fallacy of the ecological fallacy: The potential misuse of a concept and the consequences. *American Journal of Public Health* 84(5):819–824.

Scott, M.S. and S.L. Cutter. 1997. Using relative risk indicators to disclose toxic hazard information to communities. *Cartography and Geographic Information Systems* 24(3):158–171.

Selvin, H.C. 1958. Durkheim's suicide and problems of empirical research. *American Journal of Sociology* 63(6):607–619.

Sexton, K., S.G. Selevan, D.K. Wagener, and J.A. Lybarger. 1992. Estimating human exposures to environmental pollutants: Availability and utility of existing databases. *Archives of Environmental Health* 47(6):398–407.

Shapiro, J. 1990. *Radiation Protection. A Guide for Scientists and Physicians.* 3rd ed. Cambridge, MA: Harvard University Press.

Shipler, D.B., B.A. Napier, W.T. Farris, and M.D. Freshley. 1996. Hanford Environmental Dose Reconstruction Project—An overview. *Health Physics* 71(4):532–544.

Souleyrette, R.R., II and S.K. Sathisan. 1994. GIS for radioactive materials transportation. *Microcomputers in Civil Engineering* 9(4):295–303.

Stallones, L., J.R. Nuckols, and J.K. Berry. 1992. Surveillance around hazardous waste sites: Geographic information systems and reproductive outcomes. *Environmental Research* 59(1):81–92.

Stockwell, J.R., J.W. Sorensen, J.W. Eckert, and E.M. Carreras. 1993. The U.S. EPA geographic information system for mapping environmental releases of Toxic Chemical Release Inventory (TRI) chemicals. *Risk Analysis* 13(2):155–164.

Susser, M. 1994a. The logic in ecological: I. The logic of analysis. *American Journal of Public Health* 84(5):825–829.

Susser, M. 1994b. The logic in ecological: II. The logic of design. *American Journal of Public Health* 84(5):830–835.

Thomas, R.D. 1995. Age-specific carcinogenesis: Environmental exposure and susceptibility. *Environmental Health Perspectives* 103(Suppl. 6):45–48.

Tim, U.S. 1995. The application of GIS in environmental health sciences: Opportunities and limitations. *Environmental Research* 71(2):75–88.

Tobias, R.A., R. Roy, C.J. Alo, and H.L. Howe. 1996. Tracking human health statistics in "Radium City." *Geo Info Systems* 6(7):50–53.

Twigg, L. 1990. Health based geographical information systems: Their potential examined in the light of existing data sources. *Social Science and Medicine* 30(1):143–155.

United Church of Christ. 1987. *Toxic Wastes and Race in the United States.* New York: Commission for Racial Justice.

Viel, J.-F. and S.T. Richardson. 1991. Adult leukemia and farm practices: An alternative approach for assessing geographical pesticide exposure. *Social Science & Medicine* 32(9):1067–1073.

Viel, J.-F. and S.T. Richardson. 1993. Lymphoma, multiple myeloma and leukaemia among French farmers in relation to pesticide exposure. *Social Science & Medicine* 37(6):771–777.

Vine, M.F., D. Degnan, and C. Hanchette. 1997. Geographic information systems: Their use in environmental epidemiologic research. *Environmental Health Perspectives* 105(6):598–605.

Wagenknecht, L.E., L.L. Perkins, G.R. Cutter, S. Sydney, G.L. Burke, T.A. Manolio, D.R. Jacobs, Jr., K.A. Liu, G.D. Friedman, and G.H. Hughes. 1990. Cigarette smoking behavior is strongly related to educational status: The CARDIA study. *Preventive Medicine* 19(2):158–169.

Waller, L.A. 1996a. Epidemiologic uses of geographic information systems. *Statistics in Epidemiology Report* 3(1):1–7.

Waller, L.A. 1996b. Geographic information systems and environmental health. *Health and Environment Digest* 9(10):85–88.

Wang, J. and Y. Xie. 1994. Application of geographical information systems to toxic chemical mapping in Lake Erie. *Environmental Technology* 15(8):701–714.

Wartenberg, D. 1992. Screening for lead exposure using a geographic information system. *Environmental Research* 59(2):310–317.

Wartenberg, D., M. Greenberg, and R. Lathrop. 1993. Identification and characterization of populations living near high-voltage transmission lines: A pilot study. *Environmental Health Perspectives* 101(7):626–632.

Wigle, D.T., R.M. Semenciw, K. Wilkins, D. Riedel, L. Ritter, H.I. Morrison, and Y. Mao. 1990. Mortality study of Canadian male farm operators: Non-Hodgkin's lymphoma mortality and agricultural practices in Saskatchewan. *Journal of the National Cancer Institute* 82(7):575–582.

Wood, D.J. and B. Gray. 1991. Toward a comprehensive theory of collaboration. *Journal of Applied Behavioral Sciences* 27(2):139–162.

Wynder, E.L. and S.D. Stellman. 1992. The "over-exposed" control group. *American Journal of Epidemiology* 135(5):459–461.

Zito, G.V. 1975. *Methodology and Meanings: Varieties of Sociological Inquiry.* New York: Praeger.

Chapter Six

Infectious Diseases and GIS

Infectious diseases continue to plague populations throughout the world. This chapter highlights applications of geographic information systems to problems of infectious disease. Studies that have focused on dracunculiasis, Lyme disease, babesiosis, encephalitis, and malaria are highlighted. For each disease the text follows a sequence that includes (1) a description of the disease and its transmission chain, (2) a snapshot of the current geographic patterns and recent statistics, and (3) a review of selected research summarizing applications of geographic information systems. Note here that the reference material for items one and two are from the Internet sites of the World Health Organization, Centers for Disease Control, and the Tick Research Laboratory (see appendix for URLs). Source materials for item three are largely drawn from the Master GIS/RS Bibliographic Resource Guide (this volume).

DRACUNCULIASIS (GUINEA WORM DISEASE)

Dracunculiasis (Guinea Worm Disease) is a water-borne disease that involves the interaction between cyclops (minute crustaceans), Guinea worms, and humans. This disease is endemic in sub-Saharan Africa north of the equator. The global incidence of dracunculiasis has declined dramatically over the last two decades. From 1992 to 1997 the global incidence of dracunculiasis has dropped from 423,000 to 72,000 cases. Of these 72,000 cases, 57% were from Sudan. The transmission cycle of dracunculiasis begins with people drinking water containing cyclops infected with larvae of Guinea worms. Digestive juices within the human gut release the larvae from the cyclops. The larvae then migrate into the abdomen to mature and mate. Female adult worms travel to subcutaneous tissue, usually within the lower leg or foot. The adult female worm ruptures through human tissue and upon submersion of infected limbs in water hundreds of thousands of larvae reenter ponds and watering holes. The cycle completes itself with cyclops feeding on the new generation of Guinea worm larvae. Eradication campaigns that have focused on providing insecticides to eliminate cyclops, on cleaning the water supply (construction of cisterns and nylon mesh filters that screen out cyclops), and on health education have been rather successful.

The Geographic Information System for the Drancunculias Eradication Programme (WHO, 1996a,b) is a joint operation between the World Health Organization and the United Nations Children's Fund (UNICEF). One specific project aim involved developing a GIS database for Zou Province, Benin (Clarke et al., 1991). Spatial coverages of settlements (derived from remote sensing data), political and administrative boundaries, hydrology, water wells and drill holes were brought together within a GIS environment. Other information such as dracunculias incidence, vital statistics, and well characteristics (type of well, condition, type of pump, dates of drilling) was linked with the spatial coverages. Future plans call for using Global Positioning System (GPS) receivers to record data in the field. These geographic databases are providing WHO, UNICEF, and participant states of the Geographic Information System for the Dracunculiasis Eradication Programme the technological infrastructure to monitor and eradicate dracunculiasis.

LYME DISEASE

Lyme disease is caused by *Borrelia bungdorferi,* a corkscrew-shaped bacterium, which is transmitted primarily by ticks (*Ixodes scapularis*). The disease cycles among the white-footed mouse, white-tailed deer, other mammals (including humans), and birds. In 1996 there were 16,455 cases of Lyme disease reported to the Centers for Disease Control (CDC). The highest rates were reported from Maryland to Maine, in Minnesota and Wisconsin, and in northern California. Early symptoms of the disease usually include fatigue, chills and fevers, headache, muscle and joint pain, swollen lymph nodes, and a characteristic skin rash called erythema migran. Other late-stage symptoms might produce arthritis, nervous system abnormalities, and heart rhythm irregularities. The disease is treatable with antibiotics; however, on rare occasions death occurs. The CDC recommends the following preventable measures for personal protection against tick bites. These include avoiding tick-infested areas (especially in May, June, and July), wearing light-colored clothing, tucking in garments, using an appropriate insect repellant on exposed skin surfaces (avoid face) and clothes, wearing a hat and long-sleeved shirt, and walking in the center of hiking trails to avoid lurking ticks.

Lyme disease is a vector-borne disease that is amenable to both geographical information systems and remote sensing techniques. Some studies have incorporated both techniques (Glass et al., 1995; Kitron and Kazmierczak, 1997), whereas other studies incorporate just remote sensing (see Chapters 7 and 8). Table 6.1 presents a synopsis of the study areas, variables, analyses, and conclusions from five Lyme disease studies using GIS functions.

There have been two approaches used to incorporate environmental variables into Lyme disease studies. The first approach is to include just variables assumed to have a high association with either tick distribution (Kitron et al., 1991) or human cases (Glass et al., 1995; Kitron and Kazmierczak, 1997). The second approach is to include several dozen environmental variables and then attempt to determine the importance of these through spatial statistics (Glass et al., 1994; Glass et al., 1995). Regardless of the approach, coverage of soil, vegetation, and water (hydrology, drainage basins) seems to be important. Thematic mapping has been used to describe patterns of selected environmental variables, tick distribu-

Table 6.1. Synopsis of Lyme Disease-GIS Applications.

Study Area	Variables	Analysis	Conclusions	Authors
Illinois	soil vegetation cover distance to water tick distribution infected and uninfected deer	plotting overlay	sandy soils, wooded vegetation, proximity to rivers; infected deer clustered around endemic areas, noninfected deer no clustering	Kitron et al., 1991
Wisconsin	vegetation (NDVI) tick distribution human cases population density	thematic mapping autocorrelation	risk map correlated with human cases and tick distribution clusters in western Wisconsin	Kitron and Kazmierczak, 1997
Rhode Island	road network land use vegetation cover hydrography tick distribution human cases population	autocorrelation	strong association among Lyme disease, nymphal, and prevalence of infected tick; plant communities not predictive	Nicholson and Mather, 1996
Maryland	41 environmental variables tick distribution	regression Thiessen polygons	associations with well-drained soil, low water tables, and sandy soils	Glass et al., 1994
Maryland	53 environmental variables land use/land cover forest distribution soils elevation geology watershed residence locations of cases and controls	regression	association with residence, forested areas, specific soils, well-drained	Glass et al., 1995

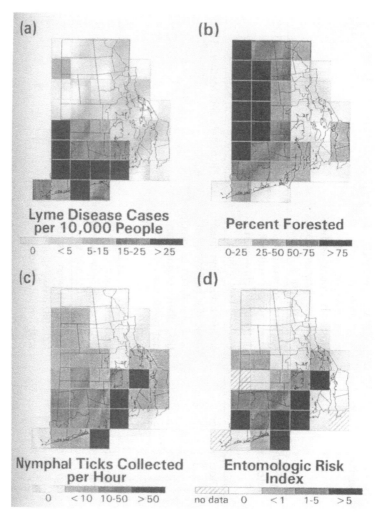

Figure 6.1. Spatial data derived from a GIS and field collections to determine Lyme disease risk in occurrence in Rhode Island. Data entered into analysis from 10 km², quadrats included: (a) number of Lyme disease cases per capita (1992–1993); (b) proportion of quadrat in forest, (c) nymphal tick densities, (d) density of uninfected nymphal ticks (entomologic risk index). *Source: Journal of Medical Entomology,* 33, M.C. Nicholson and T.N. Mather. Methods for Evaluating Lyme Disease Risks Using Geographic Information Systems and Geospatial Analysis, pp. 711–720, 1996. Reprinted with permission from the Entomological Society of America.

tions or human cases (Figure 6.1). A wide range of spatial analytic techniques has been used within the referenced literature; these include: overlay analysis (Kitron et al., 1991), regression analysis (Glass et al., 1994, 1995), kriging (Nicholson and Mather, 1996), and autocorrelation (Kitron and Kazmierczak, 1997). Several studies have produced risk maps based on adult *Ixodes scapularis* abundance per white-tailed deer (Glass et al., 1994; Figures 6.2 and 6.3), logistic regression of environmental variables (Glass et al., 1995), and the density of nymphal ticks

Figure 6.2. Distribution of 18 Thiessen polygons used to measure habitat variables in Kent County, Maryland. Number corresponds to locations listed in Table 1 (from original article). *Source: American Journal of Tropical Medicine and Hygiene,* 51, G.E. Glass et al. Predicting *Ixodes scapularis* Abundance on White-Tailed Deer Using Geographic Information Systems, pp. 538–544, 1994. Reprinted with permission from The American Society of Tropical Medicine and Hygiene.

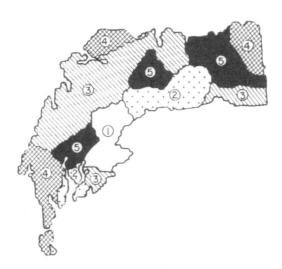

Figure 6.3. Predicted abundance of adult *Ixodes scapularis* ticks per white-tailed deer in Kent County, Maryland Grouped into five classes. Predicted abundance increases from 1 through 5 with Class 1 = 0–3 ticks, class 2 = 4–7 ticks, class 3 = 8–10 ticks, class 4 = 11–12 ticks, and class 5 = > 12 ticks. *Source: American Journal of Tropical Medicine and Hygiene,* 51, G.E. Glass et al. Predicting *Ixodes scapularis* Abundance on White-Tailed Deer Using Geographic Information Systems, pp. 538–544, 1994. Reprinted with permission from The American Society of Tropical Medicine and Hygiene.

(Figures 6.4 and 6.5). Such risk maps might stimulate hikers, campers, hunters, and other outdoor adventurers to take additional preventative measures when venturing into high-risk areas.

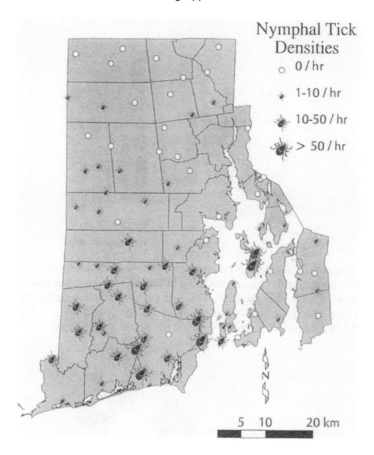

Figure 6.4. Density of nymphs as determined by flagging at 79 sample locations in 1993. *Source: Journal of Medical Entomology,* 33, M.C. Nicholson and T.N. Mather. Methods for Evaluating Lyme Disease Risks Using Geographic Information Systems and Geospatial Analysis, pp. 711–720, 1996. Reprinted with permission from the Entomological Society of America.

HUMAN BABESIOSIS

Human babesiosis is a malaria-like infection. Most cases have been reported from northeastern U. S. coastal locations such as Nantucket Island and Martha's Vineyard in Massachusetts; Block Island in Rhode Island; and Long Island and Shelter Island in New York. The etiologic agent is *Babesia microti* which circulates between blacklegged ticks and white-footed mice; however, this parasite has also been detected in the meadow vole, eastern chipmunk, Norway rat, cottontail rabbit, and short-tailed shrew. This is a rare disease among humans with just a couple hundred cases being reported over the last two decades across the United States. It is possible for blacklegged ticks (*Ixodes scapularis*) to have concurrent infections of *Babesia microti* and *Borrelia burgdorferi* (Lyme disease). Two other tick species (*Ixodes pacificus* and *Ixodes trianguliceps*) have also been shown to transmit *Babesia microti* under experimental conditions. Mather et al. (1996) were able to combine the spatial coverages of *Babesisa* occurrence, determined by capturing

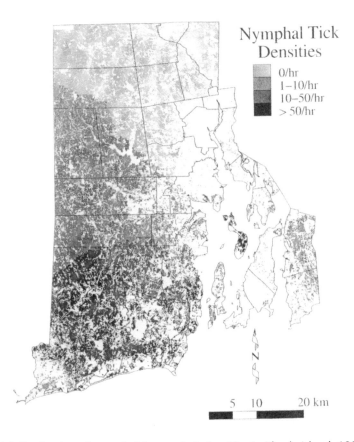

Figure 6.5. Spatial distribution of nymphal *I. scapularis* densities in Rhode Island, 1993. Continuous surface of tick densities in forested habitats was estimated by punctual kriging of point samples using a Gaussian model. *Source: Journal of Medical Entomology, 33,* M.C. Nicholson and T.N. Mather. Methods for Evaluating Lyme Disease Risks Using Geographic Information Systems and Geospatial Analysis, pp. 711–720, 1996. Reprinted with permission from the Entomological Society of America.

and testing rodents (white-footed mice) from 34 sites throughout southern Rhode Island, with the mean number of nymphal ticks found among samples from the same sites (Figures 6.6 and 6.7). Their conclusion was that *"Babesia microti* appears to have a more limited spatial distribution in Rhode Island than Lyme disease spirochetes, although both infections appear to be constrained by the distribution of *I. Scapularis"* (Mather et al., 1996, p. 868). Mather and colleagues foresee utilizing a geographic information system to predict patterns of *B. microti* among zoonotic hosts.

LACROSSE ENCEPHALITIS

LaCrosse (LAC) encephalitis is a disease caused by a virus. This disease produces fever, headache, nausea, vomiting, and lethargy in humans. LAC can also cause seizure, coma, and paralysis; however, less than 1% of clinical cases results in death. Children under 16, especially males, are more susceptible. Perhaps this

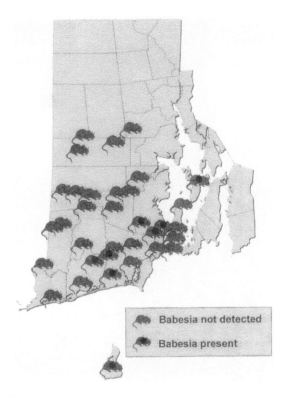

Figure. 6.6. Spatial distribution of *B. microti*-infected white-footed mouse (*P. leucopus*) populations in Rhode Island, 1994. At least 10 rodents were captured and tested at each site. *Source: Journal of Medical Entomology*, 33, T.N. Mather et al. Entomological Correlates of *Babesia microti* Prevalence in an Area Where *Ixodes scapularis* (Acari: Ixodidae) is Endemic, pp. 866–870, 1996. Reprinted with permission from the Entomological Society of America.

higher susceptibility among boys is related to their traditional preference for outdoor play and recreational activities.

LAC virus circulates within deciduous forest habitats through the interaction between the treehole mosquito (*Ades triseriatus*), chipmunks, and tree squirrels. On biting (taking a blood meal), infected mosquitoes pass on the virus to humans. From 1964–1997, 27 states reported 2,478 confirmed and probable human cases of LAC to the Centers for Disease Control. In the last few years, 1995–1997, the states with the greatest number of cases were West Virginia (166), Ohio (44), Illinois (22), Tennessee (20), Iowa (12), and Minnesota (12). Note that these states form a contiguous northwest-southwest trending region from Minnesota to West Virginia.

Kitron et al. (1997) examined the spatial distribution of LAC in Illinois using a geographic information system. Through a series of cartographic and spatial analyses zooming in from the county and down to town and exact address level, the foci of the distribution of cases from 1966 to 1995 were found to be in and around Peoria (Figures 6.8, 6.9, and 6.10). Using the Getis-Ord Gi(d) local statistic calculated over a 10 km distance around each town in Knox, Peoria, Woodford, and Tazewell Counties, ten towns were found to have significant levels of encephalitis

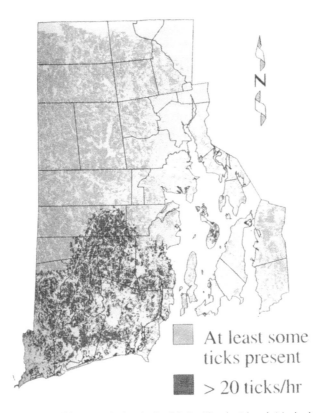

At least some
ticks present

> 20 ticks/hr

Figure 6.7. Area of potential human babesiosis risk in Rhode Island (dark shading). *Source: Journal of Medical Entomology,* 33, T.N. Mather et al. Entomological Correlates of *Babesia microti* Prevalence in an Area Where *Ixodes scapularis* (Acari: Ixodidae) is Endemic, pp. 866–870, 1996. Reprinted with permission from the Entomological Society of America.

case clustering. Eight of the ten towns having significant case clustering were within a 15-km radius of Peoria (city). Further, second-order analysis of address-matched cases found that clustering occurred within a range of 3.0 km. This rather limited range (3.0 km) suggests intervention efforts might focus around locales that juxtapose case clusters with the breeding sites of treehole mosquitoes (hard-wood-clad ravines, tire piles). Kitron et al.'s (1997) application is exceptional for two main reasons. First, examining the distribution of cases at several geographic scales prevents a myopic interpretation of data. Second, the use of spatial statistics is essential in locating and evaluating potential disease clusters. However, using almost thirty years of data on cases might obscure changing patterns of clusters over the last three decades.

MALARIA

Malaria is a disease common to the tropical areas of Africa, Asia, and Latin America with some 300 to 500 million clinical cases and 1.5 to 2.7 million deaths each year. Malaria is a disease of gargantuan proportions. The disease agent is one

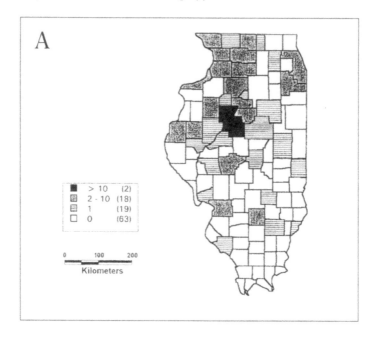

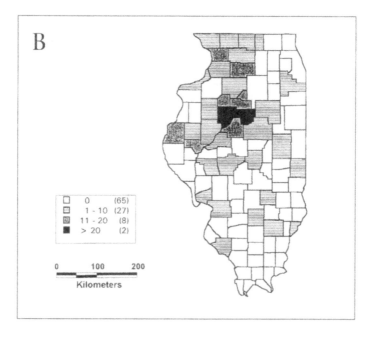

Figure 6.8. Distribution of LaCrosse encephalitis in Illinois by County, 1966–1995. A, total number of cases. B, incidence per 100,00 of county population densities based on 1990 census data. *Source: American Journal of Tropical Medicine and Hygiene,* 57, U. Kitron et al. Spatial Analysis of the Distribution of LaCrosse Encephalitis in Illinois, Using a Geographic Information System and Local and Global Spatial Statistics, pp. 469–475, 1997. Reprinted with permission The American Society of Tropical Medicine and Hygiene.

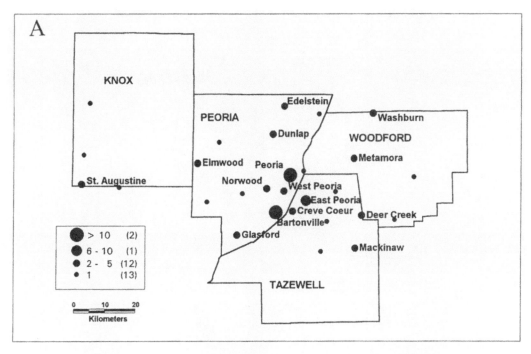

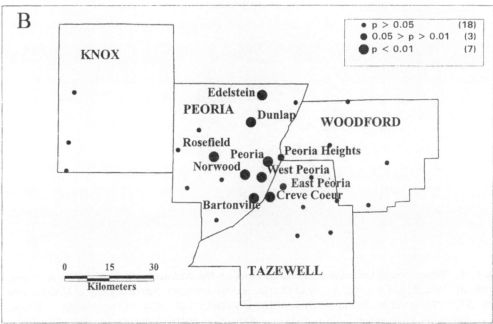

Figure 6.9. Distribution of LaCrosse encephalitis cases by town in the Peoria region, 1966–1995: A, number of cases. B., significance level of clustering of cases as measured by the Gi(d) local statistic over a distance of 10 km around each town. *Source: American Journal of Tropical Medicine and Hygiene, 57,* U. Kitron et al. Spatial Analysis of the Distribution of LaCrosse Encephalitis in Illinois, Using a Geographic Information System and Local and Global Spatial Statistics, pp. 469–475, 1997. Reprinted with permission The American Society of Tropical Medicine and Hygiene.

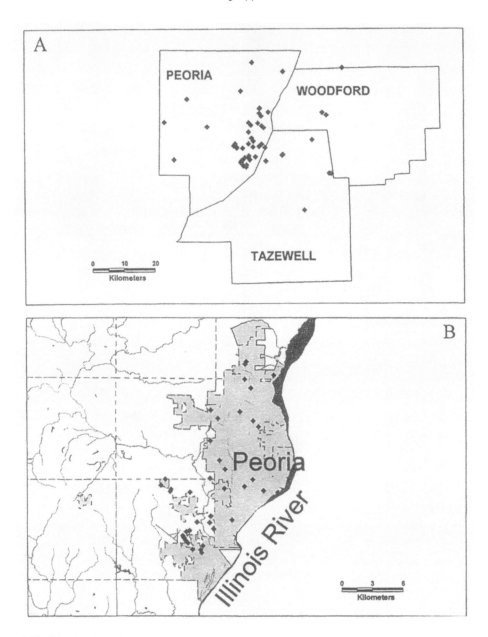

Figure 6.10. Distribution of LaCrosse encephalitis cases on the address level, 1976–1995. A, Peoria, Tazewell, and Woodford Counties. B. city of Peoria. *Source: American Journal of Tropical Medicine and Hygiene,* 57, U. Kitron et al. Spatial Analysis of the Distribution of LaCrosse Encephalitis in Illinois, Using a Geographic Information System and Local and Global Spatial Statistics, pp. 469–475, 1997. Reprinted with permission The American Society of Tropical Medicine and Hygiene.

of four protozoan parasites: *Plasmodium falcoprum, Plasmodium vivax, Plasmodium ovale,* and *Plasmodium malaria.* The disease is vectored (transmitted) to humans by the bite of female anopheline mosquitoes. Although there are several hundred species of *anopheles*, only 60 are known to transmit the parasite. The clinical symptoms of persons with malaria include fever, shivering, pain in the joints, and headache. One major concern is that parasites are developing resistance to some common drugs used to treat persons with malaria.

Since anopheline mosquitoes require a water source for breeding, geographic information systems that integrate spatial databases such as hydrology, canals and irrigation networks, ditches and other collectors of rain can assist in identifying risk areas. Malaria, like encephalitis, is a disease that can benefit from both GIS and remote sensing. Our focus here is on four studies that use GIS for malaria control and prevention. These studies were conducted in various countries: India (Sharma and Srivastava, 1997), Ethiopia (Ribeiro et al., 1996), Israel (Kitron et al., 1994), and Mozambique (Thompson et al., 1997). When one considers also the studies using remote sensing highlighted in Chapters 8 and 9, the geographical cross section of study areas is very representative of the worldwide distribution of malaria.

Most of the GIS databases (Table 6.2) constructed to analyze patterns of malaria include spatially referenced information on entomologic inoculation rate, malaria incidence or cases, and mosquito densities. Another important variable is the location of breeding sites (Kitron et al., 1994; Thompson et al., 1997). The research presents two approaches to examining infested areas. The first approach is to associate malaria incidence with elements of physical geography; that is, constructing a GIS database that juxtaposes malaria incidence or prevalence rates with soil, hydrogeomorphology, water table, water quality, and relief coverages. The idea, of course, is to identify specific locales where these elements produce habitats most suitable for mosquito populations. The second approach is to develop a GIS database that facilitates the association of mosquito breeding sites, mosquito densities, and entomologic inoculation rates (rate of mosquito bites) with demographics (age, sex, and occupation) and the built environment (house characteristics such as roof, wall, and window construction). Perhaps the incorporation of both the physical and human geographic approaches is the next logical step for future investigations.

Thompson et al. (1997) found a steep gradient in malaria prevalence with distance from breeding sites and Kitron et al. (1994) noted that localized malaria outbreaks were correlated with proximal breeding sites. Ribeiro et al. (1996) found that clusters of mosquito densities dominate the periphery or edge of a village; further, that the pattern of these clusters changed over time.

To further illustrate, note the map (Figure 6.11) of *Plasmodium falciparum* (malaria) prevalence distribution for a suburban area (Matola) of Maputo, Mozambique (Thompson et al., 1997). Matola is on the southern coastal plain of Mozambique; the southern fringe of Matola is bounded by a salt marsh that separates the town from Maputo Bay. Railways and highways enclose most of Maputo within a triangular sector. Just six of Matola's 43 districts lie between the highway and the salt marsh. Total population for the 43 districts was 21,897. *Plasmodium falciparum* prevalence rates are shown as pie graphs that indicate the proportion of each district's population infected (positive) and uninfected (negative). Note the higher prevalence rates, 6.2 times greater risk, in districts in close proximity to water bodies, canals, and other potential breeding sites.

Table 6.2. Synopsis of Malaria-GIS Applications.

Study Area	Variables	Analysis	Conclusions	Authors
Mozambique	breeding sites housing (roof, wall, window) demographics entomologic inoculation rate mosquito density	buffer zones regression proportional symbols	steep gradient in prevalence from breeding sites; target control efforts	Thompson et al., 1997
Israel	breeding sites population centers imported cases vectorial capacity flight range	distance maps	localized outbreaks are correlated with proximal breeding sites	Kitron et al., 1994
Ethiopia	mosquito densities (household)	temporal plots kriging	clustering changes over time; edge of village; focal spraying	Ribeiro et al., 1996
India	malaria incidence soil water table water quality relief hydrogeomorphology	overlay analysis mapping	control efforts ought to be location specific	Sharma and Srivastava, 1997

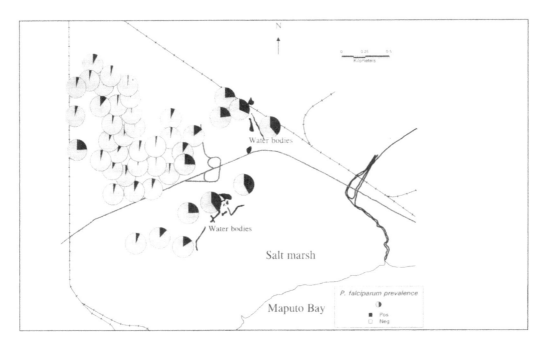

Figure 6.11. Map of the Palsmodium falciparum prevalence distribution at the first survey in December 1992. Pos = positive; Neg = negative. *Source: American Journal of Tropical Medicine and Hygiene*, 57, R. Thompson et al. The Matola Malaria Project: A Temporal and Spatial Study of Malaria Transmission and Disease in a Suburban Area of Maputo, Mozambique, pp. 550–559, 1997. Reprinted with permission from The American Society of Tropical Medicine and Hygiene.

The four studies highlighted in this section used various analytic techniques, including overlay analysis, buffers, distance measurements, proportional symbols, graphs, maps, and kriging (interpolation). It is remarkable that all these methods have led to similar conclusions. That is, controlling mosquito populations and malaria ought to be based on local rather than broad-based approaches. For example, Thompson et al. (1997) recommend a targeted approach to malaria control and prevention; and more specifically, Kitron et al. (1994) suggested a targeting of malaria breeding sites. These sentiments were echoed by Ribeiro et al. (1996) in their advocating of focal spraying and Sharma and Srivastava's (1997) recommendation that control efforts be location specific.

CONCLUSION

This chapter examined four infectious diseases whose control benefited from using geographic information systems. For dracunculiasis, GIS provided an effective method to *monitor* the eradication of this water-borne disease across vast and remote areas throughout numerous African countries. Geographic information systems were used to generate Lyme disease *risk maps* based on nymphal densities, abundance of adult *Ixodes scapularis* per white tailed deer, or a logistic regression model that incorporated environmental variables. Spatial statistics coupled with a GIS were able to *locate and evaluate* potential clusters of LaCrosse encephalitis. Finally, geographic information systems facilitated the *overlay* of social and environmental coverages with mosquito breeding sites, mosquito densities, or malaria incidence or inoculation rates. The result has shown a close juxtaposition between malaria incidence and inoculation rates with distance to water sources. One of the common conclusions from all these studies is the potential to identify localized risk areas that can either be avoided or targeted for vector control.

REFERENCES

Clarke, K.C., J.P. Osleeb, J.M. Sherry, J.P. Meert, and R.W. Larsson. 1991. The use of remote sensing and geographic information systems in UNICEF's dracunculiasis (Guinea worm) eradication effort. *Preventive Veterinary Medicine* 11:229–235.

Glass, G.E., B.S. Schwartz, J.M. Morgan, D.T. Johnson, P.M. Noy, and E. Israel. 1995. Environmental risk factors for Lyme disease identified with geographic information systems. *American Journal of Public Health* 85 (7):944–948.

Glass, G.E., F.P. Amerasinghe, J.M. Morgan, and T.W. Scott. 1994. Predicting *Ixodes scapularis* abundance on white-tailed deer using geographic information systems. *American Journal of Tropical Medicine and Hygiene* 51(5): 538–544.

Kitron, U. and J.J. Kazmierczak. 1997. Spatial analysis of the distribution of Lyme disease in Wisconsin. *American Journal of Epidemiology* 145(6):558–566.

Kitron, U., J.K. Bouseman, and C.J. Jones. 1991. Use of the ARC/INFO GIS to study the distribution of Lyme disease ticks in an Illinois county. *Preventive Veterinary Medicine* 11:243–248.

Kitron, U., H. Pener, C. Costin, L. Orshan, Z. Greenberg, and U. Shalom. 1994. Geographic information system in malaria surveillance: Mosquito breeding and imported cases in Israel, 1992. *American Journal of Tropical Medicine and Hygiene* 50(5):550–556.

Kitron, U., J. Michael, J. Swanson, and L. Haramis. 1997. Spatial analysis of the distribution of LaCrosse encephalitis in Illinois, using a geographic information system and local and global spatial statistics. *American Journal of Tropical Medicine and Hygiene* 57(5):469–475.

Mather, T.N., M.C. Nicholson, R. Hu, and N.J. Miller. 1996. Entomological correlates of *Babesia microti* prevalence in an area where *Ixodes scapularis* (Acari: Ixodidae) is endemic. *Journal of Medical Entomology* 33(5):866–870.

Nicholson, M.C. and T.N. Mather. 1996. Methods for evaluating Lyme disease risks using geographic information systems and geospatial analysis. *Journal of Medical Entomology* 33(5):711–720.

Openshaw, S. 1996. Geographical information systems and tropical diseases. *Transactions of the Royal Society of Tropical Medicine and Hygiene* 90(4):337–339.

Ribeiro, J.M., F. Seulu, T. Abose, G. Kidane, and A. Teklehaimanot. 1996. Temporal and spatial distribution of anopheline mosquitos in an Ethiopian village: Implications for malaria control strategies. *Bulletin of the World Health Organization* 74(3):299–305.

Sharma, V.P. and A. Srivastava. 1997. Role of geographic information system in malaria control. *Indian Journal of Medical Research* 106:198–204.

Thompson, R., K. Begtrup, N. Cuamba, M. Dgedge, C. Mendis, A. Gamage-Mendis, S.M. Enosse, J. Barreto, R.E. Sinden, and B. Hogh. 1997. The Matola Malaria

Project: A temporal and spatial study of malaria transmission and disease in a suburban area of Maputo, Mozambique. *American Journal of Tropical Medicine and Hygiene* 57:550–559.

World Health. 1996a. Guinea worm eradication programme. *World Health* 49(3):24.

World Health. 1996b. Technology aids eradication campaign. *World Health* 49(3):28.

Chapter Seven

A Historical Perspective on the Development of Remotely Sensed Data as Applied to Medical Geography

How much is a picture worth? Our interpretation of the world is based upon metaphorical descriptions of images recorded by the senses. Human social development has historically been especially dependent upon the visual interpretation or conceptualization of the world; western society in particular tends to regard vision as the most important and knowledge-dependent sense. The process of seeing creates, stores, and interprets thousands of images daily. As society has advanced, so has the ability to interpret visual cues. In fact, humans are so competent at image interpretation that it is only when an attempt is made to replicate these capabilities using computer programs that we realize how powerful our innate pattern-recognizing abilities actually are. As the world and technology advance to produce more visually complex scenes, each image can be said to distill the meaning of thousands of words.

From the earliest hunter-gatherers, humans have used sensory images to interpret and study the local landscape. Modern society continues in the early personal tradition and adds imagery collected from platforms far from the surface being viewed. These modern scenes form pictures that permit heretofore impossible interpretations of the earth's surface and global landscape. Image data permit us to see differences over time; to measure sizes, areas, depths, and heights; and, in general, to acquire information that is difficult to acquire by other means. The development of new imaging technologies has promoted the development of specialized knowledge. Modern image interpretation skills are important because remotely sensed images have qualities that differ from those we encounter in everyday experience: image presentation, unfamiliar scales and resolutions, overhead or vertical views, and data collected from nonvisual regions of the electromagnetic spectrum. Many interrelated processes form remotely sensed images. An isolated focus on any single component produces a fragmented picture.

Modern remote sensing is defined as the art and science of acquiring information about an object without direct contact. The etiology of this definition reflects the combination of many changes in our historical interpretation of knowledge and the pursuit of truth. As such, remote sensing is an avenue, a reflection, of western ideology. Medical geography is the application of geographical techniques towards the resolution of health issues. The most significant portion of remote sensing applicable to medical geography "is rooted in the idea that disease-causing microbes, or the infected insects and other creatures that transmit these microorganisms to people or animals, normally reside in identifiable environments. Landscape epidemiology, as the theory is known, holds that researchers can therefore use features of the landscape to identify specific areas where the risk of transmitting these diseases is greatest" (Travis, 1997, p. 72). The broader use of medical geography techniques is covered in various chapters within this book. This chapter specifically explores the application of remotely sensed data and data manipulation techniques towards providing insight and basic information necessary to the successful resolution of health-related problems. The integration of new technologies within medical geography is clearly dependent upon the acceptance and training of individuals prepared to manage alternative and innovative data sources.

CHAPTER ORGANIZATION

This chapter is the first part of a two-chapter sequence. It is intended that this first chapter provide the framework to enable the layperson to act as an informed reader of the body of medical geography literature utilizing remotely sensed data. As such, it contains a brief history of remote sensing and introduces the basic vocabulary. The development of the technology of remote sensing parallels the use of the data within medical geography and helps to predict the direction of the discipline within the context of future applications. Chapter 8 is a detailed look at the application of remotely sensed data within the existing body of medical geography literature. The respective authors' use of the data is presented contextually in order to best explain the various techniques and not only to promote general comprehension of the remote sensing vocabulary but also to inspire ideas about how the data may be used in alternative case studies. Chapter 8 includes a number of technique-specific insets. These insets are designed to be more in-depth evaluations and discussions of the various methods used by the medical geography community when applying remotely sensed data. Chapter 8 also contains an overview of basic remote sensing terminology. Both chapters may be reviewed independently, but of course they are best understood within the context of the whole.

This chapter is intended to differ from the existing body of medical remote sensing literature that usually follows a disease-specific formula in describing those applications. The approach used here is application-specific rather than disease-specific in order to promote a more general understanding of the nature of the data and associated techniques applicable to a variety of diseases and disease vectors. Additional readings on the overview of remote sensing applications for medical geography applications are as follows: (1) Hugh-Jones (1989) provides an overview of remote sensing for disease vector research; (2) Jovanovic (1989) and Barinaga (1993) discuss the use of satellite imagery for disease prevention worldwide; (3) Washino and Wood (1994) highlight the use of satellite imagery in arthropod detection in tropical areas; and (4) Stephenson (1997) highlights the importance of using remotely sensed data to place disease research in the appropriate environmental context.

TECHNOLOGICAL ORIGINS OF REMOTE SENSING

By the early 1700s, people like Dr. Brook Taylor and J.H. Lambert, following the models of Leonardo da Vinci and the methodology of Descartes, wrote about optics and how the principles of perspective could be used to produce photographs. The earliest recorded images were created soon after with the daguerreotype developed in 1839 by Louis Daguerre. Cartographers, like Colonel Aimé Laussedat, realized the possible use of cameras and lighter-than-air craft in mapping and by 1849, were trying to take aerial photographs from kites. Colonel Laussedat had so many problems with his photographic kites that he reverted to terrestrial methods, but his work earned him the title of Father of Photogrammetry (the measurement of information with photography) (Hough, 1991; Avery and Berlin, 1985).

The advent of the 20th century brought the most significant technical addition to aerial photography, the aircraft. In 1909, a photographer accompanied Wilbur Wright in an aircraft in Centocelli, Italy and took the first photographs from this new platform (Hough, 1991). The development of aircraft-based platforms contributed to the development of photogrammetry. The maneuverability of the airplane provided the capability for the control of speed, altitude, and direction required for systematic use of the airborne camera. The Germans, led by scientists at Leica and Zeiss, were the early pioneers in the development of systematic photogrammetry. Published after WWI as part of a photogrammetry compendium, L. Fritz's "The Efficacy of Photogrammetry for Precision and Economy, With Special Reference to the Needs of Civil Engineers" illustrates the pre-WWI German methodology. Fritz proposed a model whereby the locational requirements of civil engineering could be met using photogrammetrical techniques, and thus served as the field survey replacement for all projects (Fritz, 1942).

As snapshots in time, images provide the raw material to assist in the derivation of complex environmental and human processes: in effect, to see patterns instead of isolated points and relationships between different distributions. The interpretation of the historical, processed through imagery, marks a unique divergent characteristic of imagery. However, it is important not to promote causality through imagery. Using photogrammetric methods, images can concisely convey information about locations, sizes, and interrelationships among objects. By their very nature, images portray spatial information that can be recognized as objects. These objects, in turn, illustrate history that can convey a different kind of meaning. World War I (1914–1918) marked the beginning of the routine acquisition of aerial photography. The war promoted the rapid development of equipment designed specifically for aerial photography (Campbell, 1996). More importantly, many people were trained in data collection, processing, and photo interpretation. Ultimately, these same people pioneered the post-war application of photography.

Willis T. Lee's (1922), "The Face of the Earth as Seen from the Air," surveyed a broad range of possible applications of aerial photography. The systematic use of oblique, as opposed to vertical, photography was promoted. The continued prevalence of traditional Davisian (process-oriented) description pushed most aerial photography applications towards evolutionary geomorphological purposes. Although the applications that Lee envisioned were achieved at a slow pace, the expression of governmental interest ensured continuity in the scientific development of the acquisition and analysis of aerial photography. However, technical difficulties typically associated with the development of new technology arose. These difficulties, in combination with more philosophical uncertainties regarding the role of remote observations in scientific inquiry, slowed the acceptance of the use of aerial photography (Hough, 1991).

After WWI, camera designs were improved and engineered specifically for use in aircraft. The post-war dissemination of German technological knowledge promoted the science of photogrammetry (Wolf, 1983). During the 1920s, the development of accurate photogrammetric instruments specifically designed for analysis of aerial photos further advanced the science toward its modern form. Following the creation of standardized tools and techniques, aerial photography was routinely applied in government programs (Campbell, 1996). Aerial photos were initially used for topographic mapping but later were incorporated into soil surveys, geologic mapping, forest surveys, and agricultural statistics. Supplementing survey data with photographic methods became worthy of serious consideration as surveyors used calibrated cameras to supplement survey data in remote or inaccessible corners of the world (Emmons, 1938). During World War II (1939–1945), the use of the electromagnetic spectrum was extended from almost exclusive emphasis on the visible spectrum (0.4 μm–0.7 μm) to other regions, most notably the infrared (0.7 μm–0.9 μm) used to detect camouflage. Knowledge of these regions of the spectrum had been developed in both basic and applied sciences during the preceding 150 years. However, during the war, application and further development of this knowledge accelerated, as did dissemination of the means to apply it. Wartime research and operational experience provided the theoretical and practical knowledge required for everyday use of the non-visible spectrum in remote sensing.

The systematic training and experience of large numbers of pilots, camera operators, and photo interpreters provided a large pool of experienced personnel who were able to transfer their skills and experience to civilian occupations after the war. The propagation of wartime technologies and historical methodology promoted the use of aerial photography. In part, the combination of the Davisian model and military discipline promoted the rigorous application of aerial photography and may have contributed to the continuity of use (Avery and Berlin, 1985). A significant development in the civilian sphere was the work of Robert Colwell. Colwell's research, published in 1956, utilized color infrared film to identify small-grain cereal crops and predict their diseases. Colwell's work, while important in the development of peaceful applications of photo interpretation, also illustrates an important shift in the nature of research. Colwell followed the systematic evaluation outlined by Descartes while attempting to build predictive models. The concurrent rise of positivism (the philosophy of knowledge requiring theories to be built through rigorous and repeatable hypothesis testing) in geography also illustrates this discipline-independent transition (Shaefer, 1953). The post-war era saw the continuation of trends set in motion by wartime research. On one hand, established capabilities found their way into civilian applications. At the same time, the beginnings of the Cold War created the environment for further development of reconnaissance techniques. Defense secrets slowly followed the "swords to plowshares" progression as they were replaced by more sophisticated methods. The need for improved intelligence of the closed societies of the eastern bloc led to the development of high altitude balloons and second generation cameras. The first such program, GENETRIX, proved to be a failure, but the camera technologies developed became the basis for the first satellite cameras (Hough, 1991).

The motivation provided by the Soviet's successful launch of Sputnik, combined with the Gary Powers U-2 embarrassment, furthered the development of a U.S.-designed system for satellite based image collection. Further study suggested three possible methods for the collection of satellite based images. One option was to take photographs with a camera and recover the film. This idea was rejected due to the impracticality of the collection and the short life span of the satellite. The second method used conventional cameras and film combinations to take the images while transmitting the data via television. Unfortunately, this procedure ran into the ever-present data transfer rate limitations. The third, and most practical, option was to use a television camera in space and use the same carrier wave to broadcast the images back to the ground. While this method worked from a technological point of view, it failed to meet the required spatial resolution requirements with only a best possible 33-meter instantaneous field of view. The CCD (charged coupled device) was first discovered in the 1960s by Bell Labs and offered better resolution and flexibility than the conventional TV tube. However, the TV technology proved to be more easily applied and its success did lead to the development of the TIROS class of weather satellites (Hough, 1991; Jensen, 1996).

The rapid data delivery and synoptic coverage of these weather satellites provided the first tangible benefit to the public (Hough, 1991; Campbell, 1996). It was in this context that the name "remote sensing" was first used. Evelyn Pruit, a scientist working for the U.S. Navy's Office of Naval Research, coined this term when she recognized that the term "aerial photography" no longer accurately described the many forms of imagery collected using radiation outside the visible region of the spectrum (Jensen, 1996). The release of TIROS technology to the public domain promoted the dissemination of military remote sensing technology. This technology led to the creation of instruments designed to image far outside the normal spectrum of aerial photography. The new instruments and spectrum created a wealth of remotely sensed data and spurred on the development of socially responsible applications. Early in the 1960s, the U.S. National Aeronautics and Space Administration (NASA) established a research program in remote sensing. The funding and intellectual support provided by NASA contributed to the development of applications and a burgeoning body of literature. During this same period, a committee from the United States National Academy of Sciences (NAS) studied opportunities for the application of remote sensing in the fields of agriculture and forestry. In 1970, the NAS reported the results of its work in a document that outlined many of the opportunities offered by this emerging field of inquiry (IRC, 1970).

The late 1960s and early 1970s saw the increased use of remotely sensed data applications targeted towards the improvement of society or at least assisting in the basic investigation of marginalized groups. Steiner (1966) used aerial photographs to model urban expansion in Los Angeles. Parsons and Bowen (1966) used aerial photos to analyze the influence of regional topographical characteristics in the San Jorge River valley on the local farmers. Seavoly (1973) discussed the effects of population pressures on slash and burn agricultural methods. Aerial photographs helped identify target regions and plant community succession regimes. Cline (1970) published an overview article "New Eyes for Epidemiologists: Aerial Photography and Other Remote Sensing techniques" as the first extensive review of the epidemiological applications of remotely sensed data. Cline's call for an increase in the use of remotely sensed data in medical geography coincided with concerted government programs to that end. The Health Applications Office (HAO), was set up by NASA (1970–1976) as a new arm of NASA able to investigate diseases in near real time. After 6 years of publishing and report generation, NASA closed the office. Talk at that time revolved around the use of aerospace technology for health, specifically capability and responsibility. NASA has the capability to support the collection of data and the basic infrastructure of equipment. However, the agency concluded that they didn't have either the time or the budget to conduct earthbound research. NASA was fully willing to support the health mission if the funding was provided from elsewhere (e.g., CDC, the Centers for Disease Control and Prevention). It appears that a lack of information and understanding of the possibilities of remote sensing precluded the management at any existing health agency from funding such activities (Barnes, 1991).

As more and more photos from manned space missions were made public, scientists envisioned more civilian applications. As a result, the United States in 1967 initiated the Earth Resources Technology Satellites Program (ERTS). The first satellite, ERTS-1, was launched in 1972, on board a Nimbus weather satellite platform. ERTS-1, also known as Landsat-1, was the first satellite specifically designed to image the terrestrial earth with systematic and repetitive coverage (Jensen, 1996). The spectral regions in the mid and far infrared exceeded the technical capabilities of contemporary film/filter combinations. Colvocoresses (1975) evaluated the cartographic suitability of this new data type. He found that the data were perfectly capable of meeting accuracy requirements for global mapping at 1:250,000. He also strongly argued for the systematic and uninterrupted collection of data. Despite a few system imperfections, the data were collected regularly, opening the door for true time series analysis of a variety of phenomena. Long-term permanent records are extremely useful for monitoring the extent, type, and location of environmental changes.

MORE RECENT DEVELOPMENTS

The routine availability of multispectral data collected globally greatly contributed to the expansion of the number of people who acquired experience and interest in analysis of satellite data. While multispectral data had been acquired previously, it had largely been confined to specialized research laboratories. The low cost and relative ease in data procurement contributed to a growing interdisciplinary interest in multispectral analysis. Landsat's second contribution was the rapid and broad expansion of uses of digital analyses (Campbell, 1996). Prior to Landsat-4, most analyses were completed visually by examining prints and transparencies of aerial images. Although Landsat data were available as prints or transparencies, they were also provided in digital form. Clarke et al. (1991, p. 229) remarked: "While data on the disease are, in most cases, both in short supply and unreliable, increasingly the tools of remote sensing and geographic information systems are becoming important components of the eradication effort.... When coupled with additional information available remotely sensed data from LANDSAT and when used in conjunction with a GIS containing digitized maps and field data collected from hand-held GPS receivers, the eradication effort has developed an epidemiological tool of potential power."

The routine availability of digital data in a standard (LGSOWG, Landsat Ground Station Operators' Working Group) format helped to create the market environment that increased the popularity of digital analysis and influenced the development of the now commonplace GIS software packages. Jensen's 1986 *Introductory Digital Image Processing* was one of the first remote sensing books promoting to a wide audience the use of imagery and image analysis via computers. The digital revolution revived the interest in aerial photographs. Digitally scanned photographs became primary products of the United States Geological Survey in the form of Digital Orthophoto Quads (DOQs). During the 1980s, scientists at the Jet Propulsion Laboratory began, with NASA support, to develop instruments that could create images of the earth at unprecedented levels of detail. These hyperspectral instruments imaged in many spectral regions. While the use of hyperspectral data is in its infancy, hyperspectral remote sensing is predicted to foster the emergence of a new generation of techniques and applications (Jensen, 1996).

Throughout the history of remote sensing, the availability of data has far outstripped the scientific community's ability to utilize that data (Haack, 1993). With the introduction of computer systems, the data utilization quandary had been repeatedly explored. "We must get away from the traditional way of dealing with much remotely sensed data. For many years now, we have captured the data from the sky, put it into an archive and hoped that someone somewhere, will later come along and make use of it" (Cracknell, 1991, p. 321). A significant contributing factor to the lack of penetration of remote sensing applications in the medical geography field has been the vast volume of data per satellite image. Many medical geographic researchers were not prepared in terms of computing resources to handle the large volume of data. For instance, a single-band, SPOT Panchromatic scene is 36 mb (megabytes), and a Landsat TM scene (multiple spectral bands) requires 270 mb of space. These initial space requirements are in addition to the space required for the development of image processing products. Estes et al. (1986) evaluated the use of artificial intelligence to solve the large data volume problem. They promoted the use and further development of expert systems technologies to better mine the information hidden within remotely sensed data.

While the LANDSAT program has dominated the commercial and scientific use of remotely sensed data, it is not the only commercial satellite image system. The satellite with the best spatial resolution is the SPOT (Satellite Pour l'Observation de la Terre) satellite launched by the French. SPOT satellites 1 though 4 collect 10-meter spatial resolution panchromatic data. Russia makes available the data from two former military satellite systems, the KFA-1000 and the MK-4. These systems are similar to the old CORONA satellite in that they are used to produce photographic products (unlike the digital images of LANDSAT and SPOT). Although they don't offer the flexibility of the French and U.S. systems, the archived Soviet satellite photos have a nominal spatial resolution of less than 10 meters. The new public high spatial resolution data promoted the reevaluation of older aerial photographic techniques and applications and spurred on the development of an entirely new set of satellite based data applications. Nizam (1996) used SPOT panchromatic data to map the urban expansion in post-war Beirut. Until recently, all of the spaceborne sensors were significantly affected by atmospheric conditions. The Synthetic Aperture RADAR (SAR) sensors carried aloft via the Space Shuttle, and more recently by the Ariane 4 launch vehicle, are generally unaffected by clouds and other weather conditions. SAR data are a measure of texture as opposed to reflectance characteristics. Tennakoon et al. (1992) recommended SAR for its cloud cover penetrating capabilities when evaluating monsoon dependent crops.

SENSOR RESOLUTIONS AND THEIR IMPORTANCE

Understanding differences in sensor capabilities is integral to choosing data best suited for a particular application. To assess the characteristics and trade-offs involved in data acquired from different sensors, it is necessary to have a working knowledge of the characteristics or resolutions by which data are judged (for a more complete discussion, see Chapter 8). The four types of resolution are spatial, spectral, radiometric, and temporal, and are each discussed below in the context of remotely sensed data applicable to medical remote sensing applications.

Spatial resolution is usually expressed in terms of how large an area on the ground is represented in a pixel. For example, Landsat TM data imagery is 30 m resolution, meaning that each pixel represents the reflectance of a 30 by 30 meter area on the ground. For aerial photography, spatial resolution is often loosely referred to via the scale of the photo. For example, 1 cm on a photo shot at a scale of 1:50,000 would represent a length of 50,000 cm or half a kilometer. When considering spatial or any other type of resolution, it is imperative to consider the purpose of the application; that is, the characteristics of the object one is trying to detect strongly affect the appropriateness of a given resolution specification. For instance, an analyst looking for mosquito-supporting dambo (waterlogged grass areas) habitat might find Landsat TM data to be of sufficient spatial resolution. However, a researcher looking for local isolated water bodies for smaller areas of mosquito breeding might miss important smaller streams using 30 m data, as Hayes et al. (1985) discovered in their work in Nebraska and South Dakota. Researchers new to remote sensing might immediately assume that one should always use the finest spatial resolution data available for any given project. However, as spatial resolution increases, so too does the amount of data and therefore the computer space, processing time, and cost associated with acquisition and processing.

NOAA AVHRR data (see Table 7.1) have a spatial resolution of 1100 m, and are therefore a cost-effective way of collecting information over very large areas (Linthicum et al., 1991; Davies et al., 1992). Landsat MSS data have a spatial resolution of 79 m, while Landsat TM data have 30 m resolution. There are other sensors with higher spatial resolution, including SPOT Multispectral (20 m) and SPOT Panchromatic (10 m). Many newer and planned sensors place a high priority on increasing spectral resolution. Active (non-optical) sensors also exhibit an array of spatial resolutions that vary depending upon mode. Because of the trade-off between scene size and spatial resolution, researchers should possess a working knowledge of which resolutions are most fit for a particular type of application. For instance, 1-kilometer data (e.g., NOAA AVHRR) are appropriate to "assess vegetation indices for states and entire countries [and] track regional events such as insect infestation..." (SPOT Image Corporation, 1998, p. 15). Data with 80-m resolution (or 79 m, as with Landsat TM) are appropriate for characterizing general vegetative health, whereas 20 to 30 meter resolution data are more appropriate for land cover classifications and delineation of large areas of vegetation, water, or soil. Smaller features, such as farm fields or small water bodies, may require 10-meter resolution.

The second type of resolution of concern in data selection is spectral resolution, which concerns the number and width of the bands of the electromagnetic (EM) spectrum recorded by the sensor. Optical sensors usually record in the visible and infrared portions of the electromagnetic spectrum; this span of the EM spectrum covered by a sensor is its spectral range. Within that range, the sensor may break up the reflected energy it senses into any number of bands. For example, a sensor might have two bands, one that records energy reflected in the visible spectrum and another that records in the infrared spectrum. Because these bands are few and because they each cover a broad range of the EM spectrum relative to research needs, they would be referred to as having low or poor spectral resolution. Newer developments in remote sensing technology have led to hyperspectral sensors with very high spectral resolution; that is, they have upwards of 200 bands covering the same portion of the EM spectrum as above, making each band very narrow. The concern with bandwidth has to do with the need to discern features on the landscape. With the exception of the perfect blackbody, every object reflects some amount of energy in various portions of the EM spectrum. The typical reflectance of an object can be quantified through a set of spectral signatures that can then be used as a type of key to recognize similar features elsewhere.

For example, suppose that an analyst were interested in locating the breeding habitat of a disease-carrying insect in order to more effectively plan insecticide applications. Suppose further that it was known that this particular insect preferred dead grass in which to breed. From established spectral signatures (via previous research), the analyst knows that both green and dead grass reflect at low levels in the blue (0.4–0.5 μm) portion of the EM spectrum and at medium levels in the lower portion of the near infrared (NIR) portion (0.7–0.8 μm). However, dead grass has a much higher reflectance in the red portion (0.6–0.7 μm) and a much lower reflectance in the upper portion of the NIR (0.8–1.1 μm). How would the analyst pick the appropriate spectral resolution? First, the ability to differentiate among visible (blue, green, red) bands would be important, although the inclusion of the blue band would not be vital since dead and green grass are not spectrally separable (not able to be distinguished) in that band. Second, it would also be important to be able to distinguish among different portions of the NIR section of the EM spectrum. Thus this analyst should look for data from a sensor that (1) has at least two NIR bands (or that the NIR band covers only the upper portion of NIR), and (2) separates the red band from other visible portions of the spectrum. Sensors without the capability to detect reflectance in the other bands (such as blue) would not be as important in this particular application. However, should the analyst need to distinguish among grass species, then higher spectral resolution data (perhaps with multiple, narrow red or NIR bands) might become necessary. Note that in this example no consideration was given to other land features that might need to be distinguished from both dead and green grass (such as mixed forests or barren areas); their inclusion might necessitate the inclusion of spectral bands not needed for the discrimination of grasses. In other words, to most efficiently and effectively select the appropriate satellite data, it helps to be familiar with the phenomenon of interest as well as the general study area. To aid in this selection, medical geographers should familiarize themselves with the basic applications of various wavelengths: (1) visible blue for differentiating soil from vegetation or locating shallow water; (2) visible green for discerning vegetation by health; (3) visible red for discriminating among vegetation species; (4) near infrared for general vegetation mapping, vegetation species discrimination, and vegetation health/phenological cycle (e.g., growth, maturation, senescence); and (5) mid-infrared for locating water/land boundaries and detecting moisture in vegetation or soil (SPOT Image Corporation, 1998, p. 17). Researchers should also remember that most often bands are used in combination, whether stacked for viewing purposes or used in combination to calculate indices (for more on vegetation indices, please see Chapter 8, this volume).

Radiometric resolution is the third type of resolution to consider in data selection, and refers to the sensitivity to brightness level. That is, how many brightness levels (in any given spectral band) can the sensor detect? Refer to the dead versus green grass example in the above paragraph. Dead and green grass both reflect energy in the red portion of the EM spectrum; simply having the red portion separate from the blue and green portions of the spectrum would not help to identify whether an area was predominantly dead or green grass. However, based on spectral reflectance characteristics, it is known that dead grass has a much higher (or "brighter") reflectance in the red band than green grass does. The ability to distinguish enough levels of brightness to detect this difference is radiometric resolution, and works hand in hand with spectral resolution. With satellite imagery, this resolution is normally referred to in bits; for example, Landsat TM data have 8 bit resolution, meaning that each band distinguishes among 256 levels of brightness. These values are averaged for each pixel and are referred to as brightness levels or, more commonly, DN values (digital numbers).

The fourth and final type of resolution is temporal, and refers to the time lag between periods of observation in the same area for a given sensor. Also referred to as return time, this type of resolution is especially important when performing change detection analysis (see Chapter 8) or when intra-annual seasonality is important for the phenomenon of interest. For instance, monitoring moisture potential during the year would require data for the same area at multiple times per year, a common application for AVHRR data due to the high temporal resolution. SPOT HRV Multispectral and Panchromatic sensors are an example of a sensor feature related to temporal resolution: they are pointable, meaning that they can capture imagery from areas not directly underneath them. This characteristic allows for the greater flexibility in return time and area covered than is normally allowed by a fixed orbit platform. Additionally, a researcher may need to assess the study area at a particular time in the crop or rain calendar, making the availability of satellite data within a narrow time frame extremely important. Temporal resolution is also directly related to an aspect of spatial resolution: while spatial resolution refers to pixel size, the spatial extent of an area is also important. As spatial resolution increases (i.e., pixels represent smaller and smaller areas on the ground), spatial extent (how many square kilometers are in a given image, which is related to swath width or the width of the area on the ground recorded by the sensor) decreases if data size limitations are held constant. But as spatial extent decreases, it will take successive visits to the same general area to build a large areal coverage, meaning that return time (temporal resolution) will need to be increased. Thus, as spatial resolution increases, spatial extent decreases causing a need for increased temporal resolution. But as will be discussed at the end of this chapter, increased spatial resolution is usually associated with decreased temporal resolution.

Another factor important in considering temporal resolution is that the preponderance of medical remote sensing applications tracing vector habitat requires data during seasons that are rainy and/or cloudy. Optical sensors cannot acquire surficial data through clouds, so there is a reduced application during most potential acquisition periods. This problem in periodic availability is especially pronounced in medical remote sensing applications. Many diseases' vectors rely upon and thrive in moist habitats (hence the use of moisture indices); not coincidentally, moisture conditions and resultant insect population explosions co-occur in cloudy conditions. Yet many of these vectors also have short observation periods when the information found can be analyzed quickly enough for intervention. A work around for this problem has been to utilize an active (i.e., non-optical and cloud-penetrating) sensor. Optical sensors generally are passive sensors in that they record reflected energy only. Active sensors, such as SAR (synthetic aperture radar, on both ERS-1 and RADARSAT), send energy toward the area of interest and record the amount and refraction of the reflected energy. Ambrosia et al. (1989) proposes using airborne SAR for the development and validation of radar scattering models that predict radar backscatter as a function of radar system and biophysical conditions. This paper reports that, generally, high backscatter should occur where either the canopy to surface or surface to canopy interaction of water-logged grass surfaces (dambo) exists, which would lead to the identification of the mosquito carrying Rift Valley fever.

RESOLUTION TRADE-OFFS

Weighing one's resolution needs can be a tedious process, even when only considering one of the four types of resolution (spatial, spectral, radiometric, and temporal). Considering multiple kinds of resolution needs simultaneously requires a prioritization of needs based on a thorough understanding of the research problem. As shown in Table 7.1, no sensor maximizes every type of resolution because of the inherent trade-offs involved in the different types of sensors. In effect, the primary limitation in all remote sensing sensor designs is the ability of the platform to send the data collected to the earth. For example, increasing spatial resolution necessitates a smaller ground coverage footprint. Landsat TM at 30 meters covers an area roughly 9 times that of SPOT 10-meter panchromatic data. This footprint decrease also impacts the temporal resolution. The Landsat repeat cycle of 16 days is significantly shorter than the 26-day cycle of SPOT. The higher spatial resolution sensors ameliorate that problem by offering off-nadir or pointable sensors. Similarly, increasing radiometric resolution also requires a larger volume data stream and leads to a decrease in one of the other resolution types. IRS-1D panchromatic data has a spatial resolution of 5.8 meters but only 6-bit radiometric resolution. Generally, increasing any type of resolution increases file sizes, data processing requirements, and costs. Therefore minimum resolution requirements should be set and prioritized as early in a project as is possible.

A review of the technical literature is a good place to start to see what other researchers have found to be appropriate sensors for varying applications. For example, if vegetation biomass (often used to determine crop stage) is an important factor in determining vector habitat, Landsat TM, SPOT, and ERS-1 (microwave) are all appropriate sensors to consider. For surface temperature, however, Landsat TM and NOAA AVHRR are sensors that have been successfully used before and whose resolution characteristics are appropriate to the application. There are also often shortcuts to weeding out or choosing different sensors based on resolution characteristics. SPOT Multispectral and Panchromatic are known for their high spatial resolution, whereas AVIRIS (airborne visible/infrared imaging spectrometer) is known for its hyperspectral capabilities. AVHRR data has superior temporal resolution but at the cost of spatial resolution (Huh, 1991). Landsat TM data have good spectral resolution (unlike most sensors, Landsat TM sensors capture spectral information in the blue band and have four infrared bands), but at the cost of slightly decreased spatial and temporal resolution (Brady, 1991). And sometimes the desired data, even if they do hypothetically exist (i.e., there is a sensor with those characteristics), are not available: satellites have a limited life span and do break down, clouds accumulate, and orbits aren't always maintained for the convenience of every end-user. Ultimately, the most important trade-off in the resolution conflict is cost: hyperspectral, high spatial resolution data can be captured via custom-flown airborne sensors in areas and at times defined by a user, but few users can afford the cost (Kingman, 1989).

Table 7.1. Resolutions and Platform/Sensor Characteristics.

Resolutions			Platform/Sensor Characteristics				
Spectral (bands)	Spectral (bandwidth in μm)	Spatial (m)	Radiometric (bits)	Temporal (days)	Off Nadir Viewing	Altitude (km)	Swath Width (km)
Landsat Multispectral Scanner (MSS)							
Band 1/4[a]	0.50–0.60	79	6–8	18	No	917	180
Band 2/5	0.60–0.70						
Band 3/6	0.70–0.80						
Band 4/7	0.80–1.10						
Band 8	10.40–12.60	240					
Landsat Thematic Mapper (TM)							
Band 1	0.45–0.52	30	8	16	No	705	185
Band 2	0.52–0.60						
Band 3	0.63–0.69						
Band 4	0.76–0.90						
Band 5	1.55–1.75						
Band 6	10.40–12.50	120					
Band 7	2.08–2.35	30					
NOAA Advanced Very High Resolution Radiometer (AVHRR)							
Band 1	0.58–0.68	1100	8	1	No	845 or 861	2700
Band 2	0.725–1.10						
Band 3	3.55–3.93						
Band 4	10.30–11.30						
Band 5	11.50–12.50						

Table 7.1. Resolutions and Platform/Sensor Characteristics. (Continued)

Resolutions			Platform/Sensor Characteristics				
Spectral (bands)	Spectral (bandwidth in μm)	Spatial (m)	Radiometric (bits)	Temporal (days)	Off Nadir Viewing	Altitude (km)	Swath Width (km)
NASA Thermal Infrared Multispectral Scanner (TIMS)							
Band 1	8.20–8.60	Variable	8	Variable	No	Variable	Variable
Band 2	8.60–9.00						
Band 3	9.00–9.40						
Band 4	9.40–10.20						
Band 5	10.20–11.20						
Band 6	11.20–12.20						
SPOT High Resolution Visible Sensor System (HRV)							
Multispectral Mode (XS)							
Band 1	0.50–0.59	20	8	Variable	Yes	832	60
Band 2	0.61–0.68						
Band 3	0.79–0.89						
Panchromatic Mode (Pan)							
Band 1	0.51–0.73	10	8	Variable	Yes	832	60
Indian Remote Sensing, Linear Imaging Self Scanning Camera III (LISS III)							
IRS-1C and IRS-1D, Multispectral							
Band 2	0.52–0.59	23	7	Variable	Yes	817	142
Band 3	0.62–0.68						
Band 4	0.77–0.86						
Band 5	1.55–1.70	70					

IRS-1C and IRS-1D, Panchromatic

Band 1	0.50–0.75	5.8	6	Variable	Yes	817	70

European Remote Sensing Satellite (ERSO1), Active Microwave Instrument (AMI), Synthetic Aperture Radar (SAR)

Image Mode	5.3 GHz C-Band	≤26.3 × 30	nominally 16	35	1[b]	785	100

RADARSAT Synthetic Aperture Radar (SAR)

Standard	5.3 GHz C-Band	25[c]	variable, 8–32	4–6	7	793–821	
Fine		8			10		45
Wide 1		48–30 × 28			3		165
Wide 2		32–25 × 28			3		150
ScanSAR N		50			2		305
ScanSAR W		100			1		510
Extended H		25			6		75
Extended L		35			1		170

Airborne Visible/Infrared Imaging Spectrometer (AVIRIS)

244 Bands from 0.40–2.50 μm		20	12	Variable	No	20	11

[a] Landsat MSS Bands 1–4 were originally numbered as bands 4–7 prior to Landsat 4; Band 8 existed only on Landsat 3.

[b] Number of beam positions for active sensors.

[c] Range and azimuth resolution in meters.

CONCLUDING REMARKS

During the last three decades, data acquired by satellite-borne sensors have become available and have been applied in many environmental and regional studies (Haack, 1982; Dottavio and Dottavio, 1984). Advantages of spaceborne remote sensing are the following: (1) systematic and frequent acquisition of information for areas that are difficult to access; (2) provision of a synoptic view of large features in a manageable number of images or photographs; and (3) maintenance of a permanent record of conditions at the time of acquisition (Paul and Mascarenhas, 1981). However, it is important to remember that remotely sensed data are tools, and in and of themselves do not reveal any secrets or solve any great mysteries. "In disease control, one is faced with three questions: when will a disease problem become serious, where will it occur and how much of a problem will it be.... Because RS and GIS graphics are so attractive and seductive, it is very easy to abuse them (knowingly or unknowingly)" (Hugh-Jones, 1991, pp. 160–161). The evaluation of remotely sensed data requires the combination of disciplinary knowledge, technical competence, and a conceptual querying framework. Geography has always played a significant role in the application of remotely sensed *data*, but has only played a relatively small part in the development of remote sensing as an *applied technology*. The following chapter demonstrates the use of remote sensing in medical geography applications.

REFERENCES

Ambrosia, V.G., K.G. Linthicum, C.L. Bailey, and P. Sebesta. 1989. Modeling Rift Valley fever (RVF) disease vector habitats using active and passive remote sensing systems. In *IGARSS '89 Remote Sensing: An Economic Tool for the Nineties*, pp. 2758–2760. Vancouver: IGARSS '89 12th Canadian Symposium on Remote Sensing.

Avery, T.E. and G.L. Berlin. 1985. *Interpretation of Aerial Photographs*, 4th ed. New York: Macmillan Publishing Company.

Barinaga, M. 1993. Satellite data rocket disease control efforts into orbit. *Science* 26(5117):31–32.

Barnes, C.M. 1991. An historical perspective on the applications of remote sensing to public health. *Preventive Veterinary Medicine* 11:163–166.

Brady, J. 1991. Seeing flies from space. *Nature* 351:695.

Campbell, J.B. 1996. *Introduction to Remote Sensing*, 2nd ed. New York: The Guilford Press.

Clarke, K.C., J.P. Osleeb, J.M. Sherry, J.P. Meert, and R.W. Larsson. 1991. The use of remote sensing and geographic information systems in UNICEF's *dracunculiasis* (Guinea worm) eradication effort. *Preventive Veterinary Medicine* 11:229–236.

Cline, B.L. 1970. New eyes for epidemiologists: Aerial photography and other remote sensing techniques. *American Journal of Epidemiology* 92(2):85–89.

Colvocoresses, A.P. 1975. Evaluation of the cartographic applications of ERTS-1 imagery. *The American Cartographer* 2(1):5–18.

Colwell, R.N. 1956. Determining the prevalence of certain cereal crop diseases by means of aerial photography. *Hilgardia* 26(5):223–286.

Cracknell, A.P. 1991. Rapid remote recognition of habitat changes. *Preventive Veterinary Medicine* 11:315–323.

Davies, F.G., E. Kilelu, K.J. Linthicum, and R.G. Pegram. 1992. Patterns of Rift Valley fever activity in Zambia. *Epidemiol. Infect.* 108(1):185–191.

Dottavio, C.L. and F.D. Dottavio. 1984. Potential benefits of new satellite sensors to wetland mapping. *Photogrammetric Engineering and Remote Sensing* 50(5):599–606.

Emmons, A.B. 1938. Mapping in the Nanda Devi Basin. *Geographical Review* 28(1):59–67.

Estes, J.E., C. Sailer, and L.R. Tinney. 1986. Applications of artificial intelligence techniques to remote sensing. *The Professional Geographer* 38(2):133–140.

Fritz, L. 1942. The efficacy of photogrammetry for the precision and economy, with special reference to the needs of civil engineering. In *Photogrammetry Collected Lectures and Essays,* O. von Gruber (Ed.), Boston: American Photographic Publishing Company.

Haack, B.N. 1982. Landsat: A tool for development. *World Development* 10(10):899–909.

Haack, B.N. 1993. GECA 580 Lecture. George Mason University.

Hayes, R.O., E.L. Maxwell, C.J. Mitchell, and T.L. Woodzick. 1985. Detection, identification, and classification of mosquito larval habitats using remote sensing scanners in earth-orbiting satellites. *Bulletin of the World Health Organization* 63(2):361–374.

Hough, H. 1991. Satellite Surveillance. *Loompanics Unlimited.* Port Townsend, Washington.

Hugh-Jones, M. 1989. Applications of remote sensing to the identification of the habitats of parasites and disease vectors. *Parasitology Today* 5(8):244–251.

Hugh-Jones, M. 1991. Introductory remarks on the application of remote sensing and geographic information systems to epidemiology and disease control. *Preventive Veterinary Medicine* 11:159–162.

Huh, O.K. 1991. Limitations and capabilities of the NOAA satellite advanced very high resolution radiometer (AVHRR) for remote sensing of the Earth's surface. *Preventive Veterinary Medicine* 11:167–184.

International Research Council. 1970. Remote Sensing with Special Reference to Agriculture and Forestry. Washington, DC: National Academy of Sciences.

Jensen, J.R. 1986. *Introductory Digital Image Processing: A Remote Sensing Perspective.* Englewood Cliffs, NJ: Prentice-Hall.

Jensen, J.R. 1996. *Introductory Digital Image Processing: A Remote Sensing Perspective,* 2nd ed. Upper Saddle River, NJ: Prentice Hall.

Jovanovic, P. 1989. Satellite medicine. *World Health* Jan–Feb:18–19.

Kingman, S. 1989. Remote sensing maps out where the mosquitoes are. *New Scientist* 123(1682):38.

Lee, W.T. 1922. *The Face of the Earth as Seen from the Air: A Study in the Application of Airplane Photography to Geography.* American Geographical Society Special Publication No. 4. Washington, DC: Conde Nast Press.

Linthicum, K.J., C.L. Bailey, D.R. Angleberger, T. Cannon, T.M. Logan, P.H. Gibbs, C.J. Tucker, and J. Nickeson. 1991. Towards real-time prediction of Rift Valley fever epidemics of Africa. *Preventive Veterinary Medicine* 11:325–334.

Nizam, Y. 1996. Mapping urban growth in metropolitan Beirut. In *Raster Imagery in Geographic Information Systems,* S.A. Morain and S.L. Baros (Eds.), Santa Fe, New Mexico: Onword Press.

Parsons, J.J. and W.A. Bowen. 1966. Ancient ridged fields of the San Jorge River Floodplain, Colombia. *The Geographical Review* 61(3):317–343.

Paul, C.K. and A.F. Mascarenhas. 1981. Remote sensing in development. *Science* 214(451):139–145.

SPOT Image Corporation. 1998. *Satellite Imagery: An Objective Guide.* Reston, VA: SPOT Image Corporation.

Seavoly, R.E. 1973. The shading cycle in shifting cultivation. *Annals of the Association of American Geographers* 63(4):522–528.

Shaefer, F.K. 1953. Exceptionalism in geography: A methodological examination. *Annals of the Association of American Geographers* 43(3):226–249.

Steiner, R. 1966. Reserved lands and the supply of space for the southern California metropolis. *Geographical Review* 61(3):344–362.

Stephenson, J. 1997. Ecological monitoring helps researchers study disease in environmental context. *JAMA* 278(3):189–191.

Tennakoon, S.B., V.V.N. Murty, and A. Eiumnoh. 1992. Estimation of cropped area and grain yield of rice using remote sensing data. *International Journal of Remote Sensing* 13(3):427–439.

Travis, J. 1997. Spying diseases from the sky: Satellite data may predict where infectious microbes will strike. *Science News* 152(5):72–73.

Washino, R.K. and B.L. Wood. 1994. Application of remote sensing to arthropod vector surveillance and control. *American Journal of Tropical Medicine and Hygiene* 50(6):133–144.

Wolf, P.R. 1983. *Elements of Photogrammetry: With Air Photo Interpretation and Remote Sensing,* 2nd ed. New York: McGraw-Hill, Inc.

Chapter Eight

The Integration of Remote Sensing and Medical Geography: Process and Application

Remote sensing is the process of collecting data about objects or landscape features without coming into direct physical contact with them. Most remote sensing is performed from orbital or suborbital platforms using instruments designed to measure electromagnetic radiation reflected or emitted from the earth's surface. The most common form of remote sensing is human vision. Photography is also a common and easily understandable form of capturing remotely sensed data. The placement of a camera onto an airplane vastly improves the usability of the data for scientific applications, and for our purposes, also introduces the concept of platform. The human eye is carried within the platform of the body. The location and orientation of the sensor, the eye itself, is determined by its position within the body platform and the direction or orientation is controlled by the movement of the head and body as a whole. It is important to recognize the inherent difference between sensor and platform. The human eye is an example of a passive sensor. It is unable to create its own energy source. A camera without a flash is a passive sensor. In a dark room no information can be recorded on the film. However, add a flash to the camera, and even in an entirely dark room, the flash creates enough light for the film to record some information. The camera with flash is an active sensor. Most earth observation satellites are passive sensors, with radar being the notable active exception.

Other sensors use other mediums such as magnetic fields and sound waves. These methods work on the same principles as electromagnetic remote sensing, but comprise a small part of the total data produced from remote sensing. Remote sensing is a technique that can be used in a wide variety of disciplines, but is not a traditional discipline or subject itself. The primary goal of remote sensing is not only the pursuit of knowledge, but also the application of any knowledge gained. Digital image processing helps further this goal by allowing a scientist to manipulate and analyze the image data produced by these remote sensors in such a way as to reveal information that may not be immediately recognizable in the original form. To understand the relationship of digital image processing to remotely sensed data, one should have a clear concept of the steps involved in the remote sensing process. The application of remotely sensed data to medical geography issues is multifaceted and requires at least a surficial understanding of each component to effectively proceed. This chapter draws heavily from medical applications of remote sensing to cover: basic vocabulary and concepts, data analysis, preprocessing, geometric corrections, enhancements, classification, change detection, and output (post-processing). Each step is important and must be addressed in any application.

RESOLUTION

The data characteristic of resolution is the single most important data quality. It is vital to understand each of the four types of resolution before deciding to use data within any application. Those already familiar with the types of resolution and the trade-offs inherent among those types may proceed; those who are not familiar with these concepts are strongly advised to refer to Chapter 7 before proceeding. The medical geographer quite often knows significantly more about any disease and disease vector than needed to meet the information requirement necessary for resolution specification. There are four types of resolution that impact the nature of the remotely sensed data. All remotely sensed data have each of these specific resolution characteristics: spatial, spectral, radiometric, and temporal. Spatial resolution is a measurement of the minimum distance between two objects that will allow them to be differentiated from one another in an image (Jensen, 1996). This is a function of sensor altitude, detector size, focal length and system configuration. For aerial photography, the spatial resolution is usually measured in resolvable line pairs per millimeter on the image (Jensen, 1996). For other sensors it is given as the pixel size, or dimensions (in meters) of the ground area which falls within the instantaneous field of view of a single detector within an array (Messina, 1997).

It is important to understand that scale and spatial resolution are not necessarily related. Remotely sensed data with any nominal spatial resolution may be displayed anywhere from 1:1200 to 1:1,000,000 ad infinitum. The spatial resolution and display scale are independent. However, there are practical limitations; for example, AVHRR (Advanced Very High Resolution Radiometer—for more information, see Table 7.1) data are collected at 1 km spatial resolution, and should not be displayed at 1:1000 simply because no additional information may be gained at that scale. Just as imagery is generally scale independent, diseases may be studied at many different scales. While historical geographers deal with vast sweeps of space and time, and each specific disease is but a small part of the overall schema, molecular biologists examine areas a fraction of an electron microscope field wide and deal with events occupying only minutes or seconds of time. Researchers often spend the whole of their careers looking at one level of the problem without grasping the whole epidemiological problem (Rogers, 1991). The selection of proper spatial resolution is the first requirement in data selection. For mapping purposes at any given scale, one necessarily chooses a source of imagery in which the features to be mapped are significantly larger than the limit of the resolution of the data (Cracknell, 1991). Hayes et al. (1985) used MSS data to identify specific aquatic habitats and target vegetation. They remarked on the importance of selecting the data with the proper spatial resolution to successfully identify landscape elements (Plate 1).

Sensors also are unique with regard to which portions of the electromagnetic spectrum they record. The electromagnetic spectrum is the extent of energy propagated through space between electric and magnetic fields whose range includes (in increasing wavelength) gamma rays, X-rays, ultraviolet, visible, infrared, microwave and VHF radiation. Different remote sensing instruments record different segments, or bands, of the electromagnetic spectrum. The number and size of the bands recorded by a sensor determine the instrument's spectral resolution. A sensor, like SPOT panchromatic, may be sensitive to a large portion of the electromagnetic spectrum but have poor spectral resolution since its sensitivity is contained within a single wide band. A hyperspectral sensor sensitive to the same portion of the electromagnetic spectrum but with many small bands would have greater spectral resolution. With higher spectral resolution both computer and analyst are able to better distinguish between scene elements. When combined with better spatial resolution, higher spectral resolution promotes order of magnitude improvements in image interpretability. Historically, the spectral resolution of sensors was predetermined in order to best identify land surface features considered most important. Spectral data, particularly the red (TM3), near infrared (TM4), and mid-infrared (TM7) regions, are sensitive to characteristics of the rice field canopy such as percent cover and leaf area, which can in turn be linked with plant filtering and mosquito larval habitat quality (Wood et al., 1991). More detailed information (e.g., greater spectral resolution) about how individual elements in a scene reflect or emit electromagnetic energy increases the probability of finding unique characteristics for a given feature on the landscape, allowing it to be distinguished from other features in the scene. For instance, in identifying villages at high risk for malaria transmission, Beck et al. (1994) successfully used remote sensing reflectance measurements within a limited number of bands (TM) to classify high probability landscapes of high probability for malaria. While seeming ideal, hyperspectral data are so information rich that it is often difficult to decide which bands to use in a particular application or test. Furthermore additional spectral bands means additional data storage and manipulation requirements. It is often the case that the availability of data far exceeds our ability to manage it (Plate 2).

Radiometric resolution refers to the sensitivity of the sensor to incoming radiation. The number of different levels of radiance a particular sensor can distinguish characterizes this form of resolution. It is easiest to think of this type of resolution with respect to panchromatic data. Data sampled in 6 or 8 bits record 64 or 256 shades of gray. This sensitivity to different signal levels will determine the total number of values that can be generated by the sensor (Jensen, 1996). Wood et al. (1991) used multispectral Daedalus (NASA airborne scanner) data but found that the radiometric quality of the thermal band made that particular band unusable. Temporal resolution refers to the amount of time it takes for a sensor to return to a previously recorded location. Seasonally dependent research and change detection require an understanding of and planning with respect to temporally dependent data collections. Most orbital remote sensing platforms will pass over the same spot at regular time intervals. These intervals were historically dependent strictly on the orbital parameters of the platform in question; today, however, many platforms carry sensors with pointing capabilities. This variable directional feature reduces apparent temporal resolution from weeks to days. Data collected on multiple dates allows the scientist to chart changes of phenomena through time. Examples include the growth of crops in various parts of the world, the expansion of urban areas, monitoring desertification, and, perhaps most common, the ever-changing weather.

By improving one or any combination of these resolutions, a scientist will increase the chance of obtaining accurate and useful remotely sensed data (Jensen, 1996). Roberts et al. (1991) considered the characterization of the preferred breeding sites by season of year as the first step in resolving the research problem with the quantification of the temporal availability of the breeding sites by category of habitat also important. Pope et al. (1992) were limited by the temporal resolution of the existing suite of satellite systems in the need for timely (in terms of seasonality) acquisitions. Their research over Kenya would have been enhanced by using a sensor with a pointable mirror like SPOT, though the availability of the data or the spectral bands limitation may have predicated the use of TM. In the vast majority of research using remote sensing as a data source, the scientists are often heard to lament over the need for better resolution. This general and inaccurate use of "resolution" invariably refers to spatial resolution. The downside to increased resolution is the need for increased storage space and more powerful data-processing tools (Jensen, 1996). An increase in spatial resolution from 10 meters to 1 meter is a 100 times increase in data volume per band. For these reasons, it is important to determine the minimum resolution requirements needed to accomplish a given task from the outset. This will avoid time unnecessarily wasted processing more data than is needed. It will also help to avoid the problem of too little data to allow completion of the task.

IN SITU DATA

Remotely sensed data are used for a surprising variety of applications. From hospital based medical remote sensing to GOES weather satellites, the range of potential scales is infinite. However, it is often necessary to collect ground truth data specifically correlated with the image data. Spatial location or georeferencing is often dependent upon accurately collected ground control points or quantitative samples. These data are combined with the remotely sensed data to assign qualitative values to broad classes or provide statistical validation. Consequently, the collection of in situ data may take many forms. From true synoptic field sampling to restricted laboratory sampling the range of acceptable scales of sampling are broad and generally dependent upon the cost and inclination of the researcher. With appropriate ground studies, remote sensing can identify and extensively map in a probabilistic manner the potential habitats of specific parasites and vectors (Hugh-Jones, 1989). Global positioning system (GPS) receivers are becoming commonplace with serviceable equipment available in the local sporting goods store. Differential GPS provides the precision to allow the accurate geolocation of the ground truth data.

DATA ANALYSIS

Remotely sensed data were first analyzed visually. Early aerial photographs were used in military campaigns with the identification of known tactical features deemed most important. By using various image processing techniques and methods an analyst may use the data to discern features invisible to the naked eye. These techniques include both visual processing techniques applied to hard copy data such as photographs or printouts and the application of digital image processing algorithms to the digital data (Jensen, 1996). The process of data visualization allows the analyst to examine data from all possible angles and to place entire images in context with their surroundings. Contextual organization in most cases helps to establish the link between abstract imagery and real places. The following sections on data analysis are broken into sections on analog (visual techniques) and digital image processing methods. Given the wide disparity between the two in terms of their respective application within medical geography, the section on digital techniques is substantially longer than the analog section.

ANALOG IMAGE PROCESSING

The first phase of any imagery application always includes a thorough visual inspection of the data. Even cursory visual analysis will reveal significant features within the imagery. The human image processing center is essentially an analog system very well tuned to identifying patterns in the landscape. Photogrammetric techniques precisely measure landscape element and provide ancillary information used in the feature identification. Table 8.1 shows the most commonly used elements of image interpretation. The choice of specific technique is often transparent to the analyst as all are used simultaneously in the interpretation process. Specific techniques can be applied as different objects or features are prioritized. An orderly approach to analog interpretation is especially important when the researcher has only a limited knowledge of the study area. This scenario is quite common in preliminary studies. For example, if an analyst has little or no knowledge of the study area depicted in an image s/he may use the pattern to distinguish between man-made and natural objects or arrangements as in the case of an orchard versus a natural forest. The texture of an object is also very useful in distinguishing objects. Different types of vegetation appear rough or smooth in a characteristic manner. Association combines general site knowledge with visual cues. For example, in many parts of the world sports fields near large buildings generally identify schools, whereas winding roads in the middle of forests and fields signify cemeteries (Avery and Berlin, 1985). Data visualization techniques combined with the concept of examining remotely sensed data in multiple bands of the electromagnetic spectrum (multispectral), on multiple dates (multitemporal), at multiple scales (multiscale) and in conjunction with other scientists

Table 8.1. Elements of Image Interpretation.

Element	Comments
Tone	lightness or darkness of a region
Texture	apparent roughness or smoothness
Shadow	shadows of buildings, trees, etc. help in feature identification and height determination
Pattern	the arrangement of individual objects into distinctive forms
Association	feature location among other objects without specific patterns
Shape	many natural and man-made features have distinctive shapes
Size	allows absolute measurement and assists in feature identification
Site	topographic position of a feature

(multidisciplinary), allow us to make a judgment not only as to what an object is, but its significance (Jensen, 1996).

DIGITAL IMAGE PROCESSING

Digital image processing is not only a step in the remote sensing process, but is itself a process consisting of several steps. It is important to remember that the ultimate goal of this process is to derive new data from an image and apply this new data towards resolving some question in the disease vector cycle. The steps taken in processing an image will vary between image types. Pope et al. (1992) used AVHRR optical imagery and airborne Synthetic Aperture Radar (SAR, an active and non-optical sensor) in their study of the Central Kenyan Rift Valley fever virus vector habitats (refer to Table 7.1). SAR and AVHRR are very different types of sensors with very different processing requirements. However, as discrete data types they merged well together and proved integral to the effective discrimination of habitat cover types. Regardless of the remotely sensed data type three basic steps need to be addressed: preprocessing, enhancement, and classification (Jensen, 1996).

PREPROCESSING

Preprocessing is necessary because digital imagery as collected by the sensor invariably includes artifacts of the collection process. Bad lines, skew due to orbital or flight path, atmospheric haze or pollution, and variability due to solar incidence all need to be recognized if not always analyzed. Quite often corrections for these issues are applied by the data vendor; it is simply important that the researcher be aware of their existence. Radiometric corrections may be applied to remove or mask bad lines and speckle as well as to improve the fidelity of the brightness value magnitudes. Gross geometric corrections are often applied to de-skew the data in order to correct for the earth's rotation under the platform while improving the fidelity of relative spatial or absolute locational aspects of image brightness values. The ancillary data collected during the time of acquisition (ephemeris data) are often used to model the hypothetically pure form of the image. Ideally, both radiometric and geometric preprocessing steps will reduce the influence of errors or inconsistencies in image brightness values that may limit one's ability to interpret or quantitatively process and analyze digital remotely sensed images. However, these variable environmental factors should only be considered true data errors when they obscure or create confusion among image brightness signals pertaining to surface cover types and conditions (Messina et al., 1998). Pope et al. (1992) did not correct for atmospheric conditions, while Wood et al. (1992) used a variety of preprocessing techniques to prepare the data for analysis. The decision to correct or not may strongly affect the researcher's ability to use techniques such as image differencing for information extraction. However, any application of preprocessing techniques should be carefully considered and the results strenuously verified, as haphazard preprocessing could introduce false confidence (Plate 3).

The sources of data error partially depend on the sensor and mode of imaging used to capture the digital image data. While certainly not exhaustive, sensors can be grouped into four general mechanical systems, each having their own characteristic sources of error: (1) scanned aerial photography (Beck et al. 1994), (2) optical scanners (Chwastek and Dworak, 1990), (3) optical linear arrays (Wood et al., 1991), and (4) side-looking radar (Pope et al., 1992). Radiometric noise generated by remote sensing instruments can take the form of random brightness deviations from electrical sources and coherent radiation interactions or more systematic variations having spatial structure or temporal persistence (Jensen, 1996).

There are five primary reasons or objectives for applying radiometric corrections to digital remotely sensed data; four of which pertain to achieving consistency in relative image brightness and one involving absolute quantification of brightness values. Relative correspondence of image brightness magnitudes may be desired for pixels: (1) within a single image (e.g., orbit segment or image frame), (2) between images (e.g., adjacent, overlapping frames), (3) between spectral band images, and (4) between image dates. The key here is that brightness value inconsistencies caused by the sensor and environmental noise factors listed above are balanced or "normalized" across and between image coverages and spectral bands. The other objective is the retrieval of surface energy properties such as spectral reflectance, albedo or surface temperature, which requires absolute radiometric processing. When using airborne Thematic Mapper data, Wood et al. (1991) incorporated sun angle corrections in the preprocessing to improve the potential accuracy of the results by minimizing differences between scenes. This research also used the sun illumination corrections to improve the consistency of the NDVI (discussed later) calculations. Geometric corrections are also very common prior to any image analysis. If any type of area, direction or distance measurements are to be made using an image, that image must have been rectified for those measurements to be accurate. Geometric rectification is a process by which points in an image are registered to corresponding points on a map or another image that has already been rectified. The goal of geometric rectification is to put image elements in their proper planimetric (x and y) positions.

Digital images collected from airborne or spaceborne sensors often contain systematic and unsystematic geometric errors. Some of these errors can be corrected by using its ephemeris data and known internal sensor distortion characteristics. Other errors can only be corrected by matching image coordinates of physical features recorded by the image to the geographic coordinates of the same features collected from a map or global positioning system (GPS). Geometric errors that can be corrected using sensor characteristics and ephemeris data include scan skew, mirror-scan velocity variance, panoramic distortion, platform velocity, and perspective geometry. Geometric transformations modify the spatial relationships between pixels in an image. The transformation consists of two basic operations: (1) a spatial transformation which defines the rearrangement of pixels on the image plane, and (2) a gray level interpolation which deals with the assignment of gray levels to pixels in the spatially transformed image.

The level 1A imagery used for most medical remote sensing applications having pixel coordinates $f(x,y)$ is transformed with a geometric model into image g with coordinates (x', y'). The transformation may be expressed as $x' = r(x,y)$ and $y' = s(x,y)$. If r and s are known analytically, it is possible to reverse transform the image. However, in the majority of cases, tie points (ground control points) are used where the location of subset of pixels from the distorted input image is precisely known on the output image (Gonzalez and Woods, 1992). All scenes are generally corrected using quadrilateral regions with the vertices of the quadrilaterals being the corresponding tie points. The procedure runs either row major or column major with identical results.

$$x' = c_1x + c_2y + c_3xy = c_4 \text{ and } y' = c_5x + c_6y + c_7xy = c_8$$

The interpolation of the DN (digital number, or brightness value for each pixel) values is separate from the spatial interpolation. Depending upon the coefficients c_i the procedure can yield non-integer values for $\hat{x}$ and $\hat{y}$. Since the distorted image g is digital, its pixel values are defined only at integer coordinates. Consequently, using non-integer values for x' and y' causes a mapping into locations for which no DN values are defined.

The technique most commonly used for the DN value interpolation is called the nearest neighbor approach or zero order interpolation. Nearest neighbor mapping is performed by mapping the integer coordinates (x, y) into fractional coordinates (x',y') by means of the above equations. The selection of the closest integer coordinate neighbor to (x',y') is assigned the gray level of the nearest neighbor to the pixel located at (x,y) (Gonzalez and Woods, 1992). Errors are introduced in the DN value mapping process (Jensen, 1996; ER-Mapper, 1995). The RMSE (root mean square error, a calculation representing the goodness of fit of the rectification) calculated should always be minimized and specified. The use of specific corner tie points found within the ephemeris data permits a high level of accuracy. However, many data types are not delivered with the appropriate ephemeris data or are over areas of few man-made features.

Elimination of the disease requires advance knowledge of the village's geography. The lack of data led to the use of remotely sensed data.... [T]he eradication effort has used two satellite based technologies to assist in locating settlements. First remotely sensed data from LANDSAT have allowed the identification of remote and small settlements in dracunculiasis-endemic areas... Using Benin national maps, air photographs and other sources, this image was georeferenced to latitude and longitude using control points common to both the map and image... Selecting the control points was not simple, since the lack of paved roads, locational errors on the maps and the heterogeneity of the image made point identification difficult... The georegistered image was classified with supervised classification using a maximum likelihood classifier.... The second use of satellite technology in the effort was the use of hand-held global positioning system (GPS) receivers in the field (Clarke et al., 1991, pp. 230–232).

In this example Clark et al. (1991) used the image to ground geocorrection method. This method is the correction of digital images to ground coordinates using ground control points collected from maps or ground GPS coordinates. It is often desirable to make sure one image is referenced with another more so than with any particular planimetric system. In this scenario, image to image geocorrection is specified, which involves matching the coordinate systems of two digital images with one image acting as a reference image.

ENHANCEMENT

There are many mathematical operations that can be used to enhance an image. Generally, they fall into two major categories: point operations and local operations. Point operations change the value of each individual pixel independent of all other pixels, while local operations change the value of individual pixels in the context of the values of neighboring pixels. Common enhancements include image reduction, image magnification, transect extraction, contrast adjustments (linear and nonlinear), band ratioing, spatial filtering, Fourier transformations, principal components analysis, and texture transformations (Jensen, 1996). The remainder of the image enhancement section will focus on the three primary techniques; indices, principal components analysis, and spatial filtering.

Indices

Measurement of vegetative characteristics and habitat in situ is costly in terms of both time and money, and often is not a practical approach. An alternative is the measurement of vegetative characteristics from remotely sensed data that allows for a more synoptic view as well as access to remote terrain. Assessment of vegetation has primarily involved Landsat TM (Thematic Mapper) and SPOT data to compose vegetation indices such as NDVI (normalized difference vegetation index), LAI (leaf area index), and PAR (amount of photosynthetically active radiation). Vegetation indices fall under the category of image enhancements, which make an image more interpretable for a particular application. Vegetation indices generally can be thought of as algorithms using spectral bands as input and yielding scores reflecting vegetation characteristics while controlling the influence of spectral variance from other features. Most vegetation indices are built upon the knowledge that healthy, green vegetation normally reflects 40–50% of the incident energy in the near infrared part of the spectrum (0.7–1.1 μm) while absorbing 80–90% of the incident energy in the visible (0.4–0.7 μm) portion of the electromagnetic spectrum (Jensen, 1996). Dry or barren soil reflects more energy in the visible portion of the spectrum than does healthy or green vegetation but less in the near infrared portion of the spectrum. Both dead and senescent (waning) vegetation have the highest reflectance in the visible spectrum but usually fall between dry soil and green vegetation in reflectance of near infrared energy. Theoretically, vegetation indices rely upon the spectral separability of dead or senescent vegetation, dry or barren soil, and healthy or green vegetation in the bands used to calculate such indices (Plate 4).

The most simple vegetation index is that of a single band; Jensen (1996) reports that historically Landsat MSS Bands 4, 5, 6, and 7 have been used to represent vegetative biomass and/or ground cover. Lillesand and Kiefer (1987) profile the use of AVHRR channels 1 and 2 in a simple vegetation index (VI), calculated as VI = Channel 2 – Channel 1, found to be sensitive to green vegetation. With this index, green or healthy vegetation tends to yield a high value due to high reflectance in the near-infrared and low visible reflectance; snow, water, and clouds tend to have negative VI scores due to having higher visible than infrared reflectance. Soil and rock, reflecting evenly in near-infrared and visible spectra, tend to have scores near zero.

One of the problems with the simple vegetation index (VI) is that it fails to account for systematic reflectance issues prevalent across bands in a given scene: for example, surface slope, surface aspect, and changing illumination conditions. Normalizing the simple vegetation index could compensate for some of those reflectance issues, and is done with such indices as NDVI (normalized difference vegetation index), computed generally as follows: (near infrared – red)/(near infrared + red). With Landsat TM data, the formula becomes (Band 4 – Band 3)/ (Band 4 + Band 3), with Band 4 representing the near infrared (0.76–0.90 μm) and Band 3 representing the red visible range (0.63–0.69 μm), although the first application of NDVI was used with MSS data. In the early 1970s, two versions of NDVI were calculated: NDVI6 = (MSS6 – MSS5)/(MSS6 + MSS5) and NDVI7 = (MSS7 – MSS5)/(MSS7 + MSS5). A few years later, a transformed vegetation index (TVI6) was created by adding a constant to NDVI6 and taking the square root, such that TVI6 = SQRT (NDVI6 + 0.5). Current versions of NDVI include (1) Landsat TM version, NDVI-TM = (TM4 – TM3)/(TM4 + TM3); (2) SPOT HRV version, NDVI-HRV = (XS3 – XS2)/(XS3 + XS2); and (3) AVHRR version, NDVI-AVHRR = (IR – red)/(IR + red). In general, there is a high amount of redundancy in the information contained by different vegetation indices, but there are nonetheless important differences to be gleaned when deciding which index is the most appropriate for the vegetation, study area, and available sensors (Plate 5).

Examples of vegetation indices applied in medical remote sensing are vast. By far the most commonly used vegetation index is NDVI. Because many disease vectors involve insects (often mosquitoes and flies), NDVI is used to approximate rainy conditions yielding high insect producing areas. Brady (1991) utilized NDVI to model the habitat of the trypanosome-carrying tsetse fly, and found that interannual as well as intrannual variance in NDVI was a good predictor of trypanosomiasis across time and space. Ambrosia et al. (1989) used NDVI to model *Aedes* mosquito habitat in conjunction with Rift Valley fever (RVF) outbreaks. A problem with NDVI was encountered in this research: in identifying dambos (flooded, low-lying areas), sisal plantations, coffee plantations, and riparian areas were misclassified as dambos or mosquito habitat. To correct for this overclassification, a vegetation classification was combined with the NDVI scores to accurately represent dambo areas of mosquito habitat. Davies et al. (1992), also researching RVF, found that AVHRR-based NDVI scores (especially 0.43 and higher) provided strong correlations between rainy seasons, dambo habitat, and RVF virus emergence in Zambia and Kenya. Linthicum et al. (1987) used NDVI to calculate their own PVAF (potential viral activity factor) and found both to be a reliable indicator of RVF in Kenya. Follow-up work (Linthicum et al., 1991) indicated the potential for real-time applications of AVHRR NDVI scores to select areas of possible outbreak that could then be more closely monitored with Landsat TM or SPOT data.

Other applications of NDVI include habitat modeling of the following: (1) western-malaria mosquito in California (Wood et al., 1991); (2) the parasite-carrying brown ear tick, responsible for East Coast fever, Corridor fever, and January Disease in cattle (Perry et al., 1991); and (3) Lyme disease in humans in Wisconsin (Kitron and Kazmierczak, 1997). Despite the success of NDVI in each of these studies, there are drawbacks to using such an index. Huh (1991) points out that NDVI (and VI as well) is not an accurate vegetation classifier below 20% vegetative cover, and as such may not be appropriate for delineating patchy vegetative areas. Jackson and Huete (1991) add the cautionary note that vegetation indices generally are not uniform but can instead be calculated from sensor voltage outputs, radiance values, reflectance values, and satellite digital numbers. Each is correct but each also yields a different vegetation index value for the same surface conditions. Thus it is important to maintain consistency in one's own applications and to read carefully comparative results from other research to see how exactly each vegetation index is calculated. However, the general advantage of vegetation indices remains clear: they provide information important for detection, classification, and identification of many types of origins and destinations of disease vectors, whether dambo mosquito habitats or remote villages in endemic areas. NDVI specifically provides a surrogate for rainfall patterns where precipitation data of appropriate extent are lacking (Cross et al., 1984, 1997).

Another index commonly used in medical remote sensing applications is the moisture index (MI). The index is calculated using Landsat TM imagery as (band 4 – band 7)/(band 4 + band 7). In order to correct for bare soil, a PVI (perpendicular vegetation index) is used as well, as in the Hugh-Jones et al. (1992) model of African bont tick habitat in Guadeloupe. In this study higher values of both MI and PVI were associated with more ticks in heterogeneous sites. Another method of representing moisture is through the modeling of temperature. Malone et al. (1994) created temperature difference (diurnal day-night surface temperature difference) images of Egypt that were found to significantly correlate with moisture areas associated with schistosomiasis risk in the Nile River area. The significance of this finding lies in the fact that moisture potential is often modeled using data from sensors such as Landsat TM that have with low temporal resolution but fairly high spatial and spectral resolution (Hugh-Jones et al., 1992). This work utilized AVHRR data which has lower spectral and spatial resolution than TM data but which has much greater temporal resolution as well as greater spatial extent. As such, diurnal (versus seasonal) variability in moisture as well as moisture variations across very large areas can be reliably assessed.

Principal Components Analysis

Principal Components Analysis (PCA) also falls under the category of image enhancements, and is performed both to add to the interpretability of an image for a particular application as well as to streamline the redundancy present in multispectral data. PCA is commonly used as a means of compressing data, and allows redundant data to be compressed into fewer bands that by definition are independent and non-correlated. This compression is referred to as a reduction in dimensionality of the data, indicating a decrease in the number of bands of data required to yield applicable results (Jensen, 1996, p. 172). In performing PCA, the first step involves assessing the correlation of two or more bands of imagery for a particular scene. With Landsat TM data, the three visible bands (TM1, TM2, and TM3) are often highly correlated with each other, as in examples in Jensen (1996) and Ambrosia et al. (1989). Ambrosia et al. found in their work on modeling the *Aedes* Rift Valley fever (RVF) mosquito vector that since TM data contains seven bands of information which are intercorrelated to some degree, a reduction of total scene/band variance was needed to reduce processing time required for clustering. In order to reduce the number of possible redundant data bands while still retaining greatest scene variance, they performed PCA on TM1 (0.45–0.52 µm), TM2 (0.52–0.60 µm), and TM3 (0.63–0.69 µm) and then used the first PC in their clustering and class identification. Thus three bands of visible spectrum information became one, and seven total bands became five: PC1, TM4, TM5, TM6, and TM7. Interestingly, Ambrosia et al. mention that they will try this process with SPOT data that may be more useful due to finer element resolution. Unfortunately, the increase in spatial resolution comes with a direct loss in spectral resolution, and it is unknown if the SPOT data will contain enough information to spectrally separate vegetation communities for *Aedes* habitat research because of waveband locations and lack of middle infrared bands (analogous to TM5 and TM7).

The second step after choosing correlated spectral bands involves the transformation of those bands in multispectral feature space, resulting in an uncorrelated multispectral dataset whose variance is ordered. Assume two highly correlated bands of pixel (brightness) values were used; due to the redundancy in those bands the brightness values are too clustered (not easily separable) on a simple graph of Band 1 versus Band 2 brightness values. PCA transforms or rotates the axes of the original data such that variance is maximized and the axes remain perpendicular to each other. The first axis, also the first PC, is assumed to contain overall scene luminance (the global trend or first order effect) whereas the other PCs represent intra-scene variance (the local trend or second order effect). Where the goal of the transformation is data reduction, generally PC1 alone will be used provided it accounts for some minimal threshold of total variance (usually above 90%), which is represented by a calculation from the eigenvalues and eigenvectors of the covariance matrix and computed as % variance explained by PC1 = (eigenvalue PC1 * 100)/sum (eigenvalues) (Plate 6).

The third step of PCA involves interpreting what these transformed axes mean for the new data (principal components). A matrix is created displaying the original bands used against the principal components, and shows the factor loadings for each combination of original band and principal component. These loadings represent the correlation between original bands and principal components; for example, a principal component with loadings of 0.93, 0.84, 0.92, and 0.27 in Landsat TM Bands 1, 2, 3, and 4, respectively, would indicate a strong visible spectrum component but a weak near infrared component. At this point decisions are made regarding how many and which principal components to use in the next phase of analysis. If seven components are created but the first three account for 95% of the variance in the data, processing time and space needs can be drastically reduced by using the three principal components rather than the seven original bands of information. In their work researching African bont tick habitat in Guadeloupe, Hugh-Jones et al. (1992) used PCA to separate habitat areas. They found that unlike some insect vectors, the bont tick occupies several distinct habitat types, including dry meadows, pond, rocky grasslands, and dry scrub. Once imagery was masked to include areas identified with tick habitat, they successfully employed PCA to separate and later identify the various "distinct" habitat types.

Spatial Filtering

Spatial filtering represents a series of mathematical techniques used to visually enhance the data. These spatial enhancements serve to improve the spatially dependent interpretability of the data. Spatial enhancement techniques focus on the concept of spatial frequency within an image and define homogeneous regions. These homogenous regions are identified by first locating edges. Similar edge effects are used to identify or classify common landscape features. Think in terms of the letters on this page. One identifies the outlines of the letters for recognition with respect to the spaces between the lines. It is the combination of both that permits character recognition. Spatial frequency broadly defines the relationship between edges and regions. For example, waterbodies with slowly varying changes in their DN values have low spatial frequency while urban areas, which vary radically among adjacent pixels, exhibit high spatial frequency. The specific techniques used in spatial filtering are not often used in medical geography; this absence of use is due not to their ineffectiveness, but in part to their complexity and some real difficulty in quantitatively evaluating the output (Plate 7).

CLASSIFICATION: INFORMATION GENERATION

Unlike analog image processing, which uses all of the elements listed in Table 8.1, digital image processing primarily relies on the various recorded radiance characteristics within the elements of each individual image pixel. The radiance values are most easily ordered and classified with respect to each other in a nonspatial framework. As software programs develop, the ability to recognize spatial variation and account for statistical violations in sample distributions is improved. Consequently, information extraction is an important dynamic area of research. It is incumbent upon the analyst to stay abreast of the rapidly changing state of the art in systems.

Expert systems and neural networks have been used to model human character recognition with the computer. Expert systems require the compilation of a large database of application specific. Expert systems based interpretations use this database to classify output. Neural networks modify themselves based upon selection rules. Once the neural network has completed the self-modification process to the satisfaction of the analyst, it is used to interpret and classify additional data sets. Multispectral classification employs spatial pattern recognition to create thematic layers from image data; groups of pixels are sorted into classes based on their brightness values in one or more bands. Classes may be comprised of known feature sets (such as a land cover map with soil, water, and vegetation classes) or may consist of unknown feature sets that are nonetheless distinguishable to the computer. Pattern recognition generally refers to enhancing data in order to discern patterns; while this process may be performed visually, currently spectral statistics are used to sort the image information and define and find patterns in that imagery. Pattern recognition consists of two phases, training and classification. Training involves defining the criteria or creating the set of decision rules used in the detection of pattern. Two types of training are used: supervised and unsupervised, although sometimes a hybrid approach between the two is referred to as a third method of training.

Supervised training is characterized by greater analyst control and works best in situations where the analyst is able to identify features in the imagery representative of each target class. These features may be identified on the image itself or with the use of data from other sources, such as aerial photography, maps, or various ground truth data. This approach requires a level of familiarity with both the landscape under study and the desired classes. By selecting pixels or an AOI (area of interest) representative of each class type, the analyst trains the computer to identify other pixels of similar characteristics. Thus the accuracy of the classification is highly dependent upon the quality and accuracy of training data. Hayes et al. (1985) used supervised, statistically based (parametric) classification to aid in identifying mosquito habitat in Nebraska and South Dakota, in a low-cost identification and inventory of larger freshwater plant communities and wetlands.

Unsupervised training, on the other hand, requires much less analyst input and is more automatic. For this reason it is sometimes referred to as clustering as it is based on the natural groups of pixels as they fall within feature space. The computer relies upon spectral statistics to discern patterns in the spectral information, although to some extent the analyst does stipulate parameters for the statistical decision rules (e.g., specifying within or between class variance thresholds). These groups may still be merged, deleted, or manipulated at a later time as easily as with supervised classification. The patterns detected by the computer may or may not correspond to classes meaningful for the landscape under study but are instead groups of pixels that are more similar spectrally to each other than they are to all other pixels. While supervised training is heavily dependent upon the analyst, unsupervised training is dependent upon the data. Identification of thematic groups occurs after classification, and the usefulness of the classification is directly dependent upon the ability of the analyst to appropriately interpret and identify the classes. However, unsupervised training is very useful when less is known about the data prior to the analysis. For example, Hugh-Jones et al. (1992) used unsupervised classification (combined with Principal Component Analysis and vegetation indices) to identify several distinct but previously unknown habitat types for the African bont tick in Guadeloupe.

Clustering generally is a nonspatial statistical process whereby all or many of the pixels in the input data are used regardless of their spatial proximity to one another. There are two primary types of clustering used: ISODATA and RGB. The ISODATA clustering method uses spectral distance sequentially, iteratively classifying pixels, redefining class rules, and classifying again so that eventually spectral distance patterns in the data are discernible. The RGB clustering method applies only to three band, eight bit data and uses three-dimensional feature space by dividing that space into sections that then define the clusters. Note that both clustering methods are based on spectral distance and not spatial contiguity. ISODATA clustering can be quite slow due to the multiple iterations, and does not take into account spatial homogeneity (spatial autocorrelation). However, with enough iterations this process will not be biased based on the default start point of clustering. With RGB clustering, the three input band limitation can be bothersome for some applications, but it is a fast processing method and is not biased based on the order of pixels analyzed. RGB clustering does not produce a signature set needed for comprehensive change detection.

Training, whether supervised or unsupervised, results in a set of spectral signatures that define each class. Parametric signatures, based on statistical parameters (covariance matrix and mean), can be generated by either supervised or unsupervised training and can then be used to train a statistical classifier. Nonparametric signatures are created only by supervised training and are based upon discrete objects (polygons); that is, pixels are assigned to a class based on whether or not they belong to the feature space as defined by the analyst (Plate 8).

There are several advantages and disadvantages to each approach. Parametric signatures assume normality and are slower than nonparametric, but are able to classify each pixel into one and only one class and are often more easily interpretable than feature space. Nonparametric signatures are useful for data that are distributed non-normally and also require less processing time; while they allow for overlap or unclassified pixels, the feature space extraction method (if used) can be difficult to interpret. Classification is often performed using a specific classification scheme, where a set of target classes is known a priori. Classification schemes should contain classes that are either integral to the research or easily discernible in the data itself. Classification schemes for various ecoregions have been popularized by researchers and provide a springboard for analysis when studies are conducted in areas similar to the one under which the classification scheme was built. Supervised classification is appropriate when the analyst needs only to identify relatively few classes, when the training sites have ground truth data available, or when clearly delineated and homogeneous regions can be identified to represent each class.

Unsupervised classification is best used when a large number of classes need to be identified, and is particularly useful when classes do not fall in spatially contiguous areas that are easily discernible. It is not uncommon to generate a large set of classes using unsupervised classification, and then to merge or reduce those classes via supervised classification (a hybrid approach). Additionally, the basic principles of supervised and unsupervised classification can be applied in sequence, through multiple iterations, or in some combination of order or area. For example, Beck et al. (1995) used a hybrid approach in Mexico to discover and then identify areas associated with malaria (primarily transitional swamp and unmanaged pasture). Lastly, just as there are parametric and nonparametric signatures, there are also parametric and nonparametric decision rules. Decision rules are arguably the most important component of classification, and involve the algorithm used in comparing pixel measurement vectors to signature sets (Table 8.2). Pixels meeting a specified criterion are then assigned to that class. The predominant types of nonparametric decision rules are parallelepiped and feature space (see below). Pixels unclassified after being put through the nonparametric decision rule are then either classified using a parametric rule or left unclassified. Pixels falling into more than one feature space object (overlap) can be decided by a parametric rule, by order, or left unclassified. The parametric rules commonly used include minimum distance and maximum likelihood.

In the parallelepiped decision rule, the pixel's values are compared to upper and lower limits of either (a) each band in the dataset, (b) a set number of standard deviations within the mean of each band, or (c) customized user limits. Parallelepiped is fast, does not assume normal distributions, and can also be used as an intermediary way of weeding out possible classes for a pixel before going to a more intensive processing procedure rule. However, parallelepipeds are rectangular in feature space and therefore pixels in the corner of that space may be improperly classified due to its inclusion despite its distance from the center. Feature space, however, has elliptical-shaped classes and therefore does not tend to misclassify "corner" observations. This decision rule does allow overlap and unclassified pixels, and may be difficult to interpret. Feature space can also be used on non-normally distributed data and is quite fast.

The minimum distance rule, a parametric decision rule, is also referred to as spectral distance and calculates the spectral distance between the measurement vector of a pixel and the mean measurement vector of a signature. There are no unclassified pixels, since every pixel is closer to one sample mean or another, and this rule is the next to fastest process (after parallelepiped). However, sometimes over-classification does occur where pixels that should not be classified into any group (spectral outliers) are classified anyway. Furthermore, this approach does not consider class variability, meaning that homogenous classes (like water) may be over-classified (errors of commission) whereas heterogeneous classes (like urban) will be under-classified (errors of omission).

Standard SPOT Panchromatic Scene:
10 Meter Spatial Resolution

Scene degraded to 100 meter
spatial resolution

Scene degraded to 30 meter
spatial resolution

Plate 1. Spatial Resolution. Note changes in discernible features as the spatial resolution is changed from 10 m to 30 m to 100 m. (Copyright CNES. Courtesy SPOT Image Corp.)

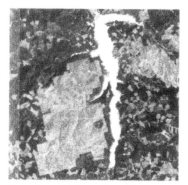

Band 1: Visible Green

Band 2: Visible Red

Band 3: Near Infra-red

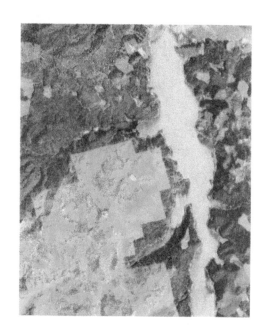

B1 + B2 + B3 = RGB
False Color Composite

Plate 2. Spectral Resolution. Shown are the green, red, and near infrared (NIR) bands of a SPOT XS scene. The composite scene composed of those three bands yields more information, but takes three times as much computer space. (Copyright CNES. Courtesy SPOT Image Corp.)

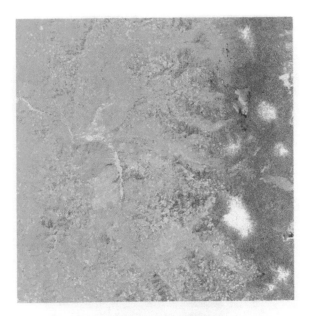

Level 1A SPOT XS Data

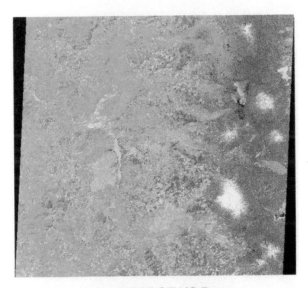

Level 1B SPOT XS Data

Plate 3. Comparison of Level 1A vs. 1B Pre-processing levels. Note obvious geometric differences. (Copyright CNES. Courtesy SPOT Image Corp.)

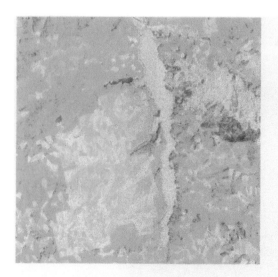

Subset of the Level 1A scene in Plate 3. Classified using an unsupervised ISODATA scheme.

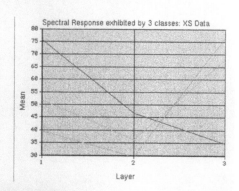

Plate 4. Classified subset of the Level 1A Scene in Plate 3. Spectral responses in the green, red, and NIR bands (layers 1, 2, and 3, respectively) are used to classify the scene. (Copyright CNES. Courtesy SPOT Image Corp.)

Original SPOT XS subset.

NDVI = (B4-B3) / (B4+3).
Normally produced in float
single output, but here is
stretched to unsigned 8 bit.

Plate 5. NDVI (Normalized Difference Vegetation Index). Shown is the same SPOT XS scene subset in Plate 4 and its corresponding NDVI. (Data courtesy of SPOT Image Corp.)

0.00340944651434601   -0.7585484484903053   0.6516077247603922
0.01929417056949305   -0.6514403135082383   -0.7584545160511099
0.9998080369033348   0.01515814069196712   0.01241451224419971

Eigenmatrix

526.6391625052033
181.4506133749329
4.234859545481283

Eigenvalues

Displayed is the image generated
by the first principal component.
This image is the same subset
used in Plate 4.

Plate 6. Principal Components Analysis of a SPOT XS scene, used to improve interpretability. Note the higher contribution of the first two Eigenvalues, suggesting that these two "bands" alone might adequately represent the spectral information. (Copyright CNES. Courtesy SPOT Image Corp.)

Standard SPOT Panchromatic Scene

3*3 High Pass Filter

5*5 Low Pass Filter

Plate 7. Spatial Filtering using High Pass and Low Pass filters. Note the changes in discernible edges of features and patterns with different filters of different kernel sizes. (Copyright CNES. Courtesy SPOT Image Corp.)

Band 1 X Band 2

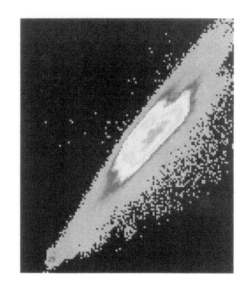

Band 3 X Band 4

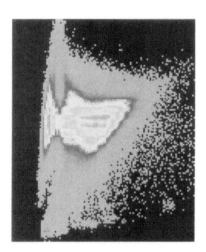

Band 2 X Band 4

Band 1 X Band 4

Plate 8. Feature Space Images: MSS Data band to band comparison. Brighter interior colors represent greater overlap between the bands, showing that here there is greater overlap between Bands 1 and 2. Conversely, Bands 2 and 4 have less overlap and would therefore be more useful for separating classes.

Table 8.2. Decision Rules and Advantages/Disadvantages.

Decision Rule	Definition	Advantage/Disadvantage
Minimum distance	Measures the distance from training region mean	Not as flexible as maximum likelihood, but better if the training region is small or of poor quality
Mahlanobias distance	Evaluates the data including the directional spread of class data in multispectral space	Similar to maximum likelihood but with fewer internal options
Maximum likelihood	Takes into account directional spread of class data in multispectral data	Recommended classifier if good quality training regions
Parallelepiped	Checks if the band value for each cell is within the minimum and maximum values of the band for the specified training region	Simple and fast qualifier not as effective as the others

Maximum likelihood (sometimes referred to as Bayesian) decision rule is based on the probability that a pixel belongs to a certain class. In its most simple form, this rule assumes these probabilities to be equal for all classes and that input data are normally distributed. This popular decision rule is one of the most accurate classifiers but only if the data are normally distributed. Class variability is taken into account via the covariance matrix, but the required processing time is therefore quite long. Additionally, this method tends to over-classify when there is a large dispersion of the pixels in a cluster or training sample. Clarke et al. (1991) successfully used maximum likelihood in their work on the Guinea worm (dracunculiasis) in Benin and Nigeria. The use of supervised, maximum likelihood classification greatly aided in identifying and locating remote and small settlements in dracunculiasis-endemic areas for intervention, with a settlement classification accuracy of 90%.

Another method of classifying the landscape involves using statistical analysis. For example, Wood et al. (1991) used discriminant analysis to test their ability to classify the landscape into areas of high and low tick larval counts using spectral reflectance and distance to field measurements. Other researchers (e.g., Wood et al., 1992) have used regression analysis and multivariate statistical models to evaluate mosquito habitat. Regression analysis explains a dependent variable(s) in terms of independent variables. Discriminant analysis, however, creates a linear combination of independent variables based on finding the greatest difference between classes of the dependent variable that were defined before data collection (as opposed to cluster analysis where classes are created based upon existing data trends—see Hugh-Jones et al., 1992). Whereas regression analysis assumes normality of the independent variable, discriminant analysis requires multivariate normality of the independent variables. While regression analysis can utilize categorical or interval data as independent variables, a categorical variable should not be used as an independent variable in discriminant analysis. Categorical data should not be used as independent variables in discriminant analysis because their non-normal distribution (often associated with noncontinuous data generally) causes model parameters estimates to be very positively biased, and a maximum likelihood estimator or logit should be used instead (Kleinbaum et al., 1988).

The other fairly serious data problem encountered in the use of statistics has to do with the spatially dependent nature of remotely sensed data. Landscape features tend to be related to those features closest to them; this phenomenon is referred to as spatial autocorrelation. For example, given a pixel with prairie land cover, the next closest pixel is more likely to be prairie than some other type of land cover. Thus spatial data are somewhat dependent upon each other, in a statistical sense. This lack of independence violates one of the most basic assumptions of regression analysis, which is that data are independent. A lack of independence causes an overestimation in the confidence of the results. There are statistical tests and corrective measures for spatial autocorrelation; for a more thorough discussion of spatial autocorrelation and other statistical applications in medical remote sensing see Chapter 2.

POST-PROCESSING

During the modeling process, it is common to find that little planning has been done to best present the output of the models. The effective transfer of knowledge and shared resources characterizes the current state of interdisciplinary research. Many of the techniques presented earlier can stand alone as informative image map output. Spatial enhancements make imagery appear more interpretable. Classifications categorize the data and provide visual cues for spatial recognition through association and quite often color. Image data itself is still simply a collection of numbers in an array. The array can be manipulated statistically and output as summary statistics and graphs. However, the tremendous number of possible output options introduces the risk of providing data commonly used in one field but not another. The output information required by a statistician is likely quite different from that desired by a geographer. The interdisciplinary nature of research and especially medical geographic research requires a thorough consideration of data output. Without sufficient consideration, it is likely that any output will be poor or at least less effective than desired and difficult to use properly, thus wasting the time and effort expended during the digital image processing phase. This chapter is not intended to serve as a map design course and as such does not provide materials related specifically to post-processing. We encourage you to review the output from the wide variety of research reviewed within this chapter and the output displayed in other chapters in this book.

CONCLUSIONS

The application of remotely sensed data to resolving health issues is a very recent innovation, but has accelerated the use and distribution of geographic techniques and remotely sensed data within the scientific community. Corbley (1999) addresses recent applications of remotely sensed data for malaria mapping in Belize, but also provides information regarding current users and federal funding agencies. The continued expansion of remote sensing applications to epidemiological issues requires the dedication and courage of scientists researching health issues. The introduction of remotely sensed data to existing programs is both expensive and time consuming. Historically, mixed results were due in part to the limited availability of data (an ever diminishing problem) and the lack of experience of the researchers (also diminishing). Fortunately, interdisciplinary research centers combining the skills of diverse groups have effectively reduced the risk of expensive mistakes and dramatically increased the use of remotely sensed data within the medical geography research community. The rapid increase is evidenced by the lengthy list of articles reviewed for this chapter. The authors highly recommend the use of remotely sensed data. Start small. Use limited numbers of scenes and types. Eventually, your confidence and ability will promote more complex image processing models and will enhance your research. Remotely sensed data are available, easy to use, and effective. We encourage you to make the leap.

REFERENCES

Ambrosia, V.G., K.G. Linthicum, C.L. Bailey, and P. Sebesta. 1989. Modeling Rift Valley fever (RVF) disease vector habitats using active and passive remote sensing systems. In *IGARSS '89 Remote Sensing: An Economic Tool for the Nineties*, pp. 2758–2760. Vancouver: IGARSS '89 12th Canadian Symposium on Remote Sensing.

Avery, T.E. and G.L. Berlin. 1985. *Interpretation of Aerial Photographs*, 4th ed. New York: Macmillan.

Beck, L.R., M.H. Rodriquez, S.W. Dister, A.D. Rodriguez, E. Rejmankova, A. Ulloa, R.A. Mesa, D.R. Roberts, J.F. Paris, M.A. Spanner, R.K. Washino, C. Hacker, and L.J. Legters. 1994. Remote sensing as a landscape epidemiologic tool to identify villages at high risk for malaria transmission. *American Journal of Tropical Medicine and Hygiene* 51:271–280.

Beck, L.R., B.L. Wood, and S.W. Dister. 1995. Remote sensing and GIS: New tools for mapping human health. *Geo Info Systems* (September):32–37.

Brady, J. 1991. Seeing Flies from Space. *Nature* 351:695.

Chwastek, J. and T.Z. Dworak. 1990. Satellite Remote Sensing of Industrial Air Pollution in the Cracow Special Protected Area. *JEPTO* 10(6):288–289.

Clarke, K.C., J.P. Osleeb, J.M. Sherry, J.P. Meert, and R.W. Larsson. 1991. The use of remote sensing and geographic information systems in UNICEF's dracunculiasis (Guinea worm) eradication effort. *Preventive Veterinary Medicine* 11:229–235.

Cracknell, A.P. 1991. Rapid remote recognition of habitat changes. *Preventive Veterinary Medicine* 11:315–323.

Corbley, K.P. 1999. Identifying villages at risk of malaria spread. *Geo Info Systems*. 9(1):34–38.

Cross, E.R., C. Sheffield, R. Perrine, and G. Pazzaglia. 1984. Predicting areas endemic for Schistosomiasis using weather variables and a Landsat data base. *Military Medicine* 149:542–544.

Cross, E., C.J. Tucker, and K.C. Hyams. 1997. The use of AVHRR and weather data to detect the seasonal and geographic occurrence of *Phebotomus papatasi* in Southwest Asia. In *Proceedings of the International Symposium on Computer Mapping in Epidemiology and Environmental Health*, R.T. Aangeenbrug, P.E. Leaverton, T.J. Mason, and G.A. Tobin (Eds.), pp. 24–26. Alexandria, VA: World Computer Graphics Foundation.

Davies, F.G., E. Kilelu, K.J. Linthicum, and R.G. Pegram. 1992. Patterns of Rift Valley fever activity in Zambia. *Epidemiology and Infection* 108:185–191.

ER-Mapper. 1995. *ER-Mapper 5.0 Reference*. Earth Resource Mapping Pty. Ltd.

Gonzalez, R.C. and R.E. Woods. 1992. *Digital Image Processing*. Reading, MA: Addison-Wesley Publishing.

Hayes, R.O., E.L. Maxwell, C.J. Mitchell, and T.L. Woodzick. 1985. Detection, identification, and classification of mosquito larval habitats using remote sensing scanners in earth-orbiting satellites. *Bulletin of the World Health Organization* 63(2):361–374.

Hugh-Jones, M. 1989. Applications of remote sensing to the identification of the habitats of parasites and disease vectors. *Parasitology Today* 5(8):244–251.

Hugh-Jones, U., N. Barre, G. Nelson, K. Wehnes, J. Warner, J. Gavin, and G. Garris. 1992. Landsat-TM identification of *Amblyomma variegatum* (Acari: Ixodidae) habitats in Guadeloupe. *Remote Sensing of Environment* 40:43–55.

Huh, O.K. 1991. Limitations and capabilities of the NOAA satellite advanced very high resolution radiometer (AVHRR) for remote sensing of the Earth's surface. *Preventive Veterinary Medicine* 11:167–184.

Jackson, R.D. and A.R. Huete. 1991. Interpreting vegetation indices. *Preventive Veterinary Medicine* 11:185–200.

Jensen, J. 1996. *Introductory Digital Image Processing: A Remote Sensing Perspective*. New Jersey: Prentice Hall.

Kitron, U. and J.J. Kazmierczak. 1997. Spatial analysis of the distribution of Lyme disease in Wisconsin. *American Journal of Epidemiology* 145(6):558–566.

Kleinbaum, D.G., L.L. Kupper, K.E. Muller, and A. Nizam. 1998. *Applied Regression Analysis and Multivariable Methods*, 3rd ed. Pacific Grove, CA: Duxbury.

Lillesand, T.M. and R.W. Kiefer. 1987. *Remote Sensing and Image Interpretation, 2nd ed.* New York: John Wiley & Sons.

Linthicum, K.J., C.L. Bailey, F.G. Davies, and C.J. Tucker. 1987. Detection of Rift Valley fever viral activity in Kenya by satellite remote sensing imagery. *Science* 235(4796):1656–1659.

Linthicum, K.J., C.L. Bailey, D.R. Angleberger, T. Cannon, T.M. Logan, P.H. Gibbs, C.J. Tucker, and J. Nickeson. 1991. Towards real-time prediction of Rift Valley fever epidemics of Africa. *Preventive Veterinary Medicine* 11:325–334.

Malone, J.B., O.K. Huh, D.P. Fehler, P.A. Wilson, D.E. Wilensky, R.A. Holmes, and A.I. Elmagdoub. 1994. Temperature data from satellite imagery and the distribution of schistosomiasis in Egypt. *American Journal of Tropical Medicine and Hygiene* 50(6):714–722.

Messina, J.P., K. Crews-Meyer, and G. Valdivia. 1998. The evaluation of preprocessing levels and classification techniques towards the better discrimination of suspended particulates. *Proceedings of the 1998 ASPRS Conference*.

Messina, J.P. 1997. Ephemeris data corrections as applied to SPOT data for marine applications. *Backscatter* 8(2):22–25.

Perry, B.D., R. Kruska, K. Kundert, P. Lessard, and R.A.I. Norval. 1991. Estimating the distribution and abundance of *Rhipicephalus appendiculatus* in Africa. *Preventive Veterinary Medicine* 11:261–268.

Pope, K.O., E.J. Sheffner, K.J. Linthicum, C.L. Bailey, T.M. Logan, E.S. Kasischke, K. Birney, A.R. Njogu, and C.R. Roberts. 1992. Identification of central Kenyan Rift Valley fever virus habitats with Landsat TM and evaluation of their flooding status with airborne imaging radar. *Remote Sensing of Environment* 40:185–196.

Roberts, D., H. Savage, L. Legters, M. Rodriguez, A. Rodriguez-Ramirez, E. Rejmankova, K. Pope, B. Wood, and J. Salute. 1991. Overview of field studies for tile application of remote sensing to tile study of malaria transmission in Tapachulah, Mexico. *Preventive Veterinary Medicine* 11:269–276.

Rogers, D.J. 1991. Satellite Imagery, tsetse and trypanosomiasis in Africa. *Preventive Veterinary Medicine,* 11:201–220.

Wood, B., R. Washino, L. Beck, K. Hibbard, M. Pitcairn, D. Roberts, E. Rejmankova, J. Paris, C. Hacker, J. Salute, P. Sebesta, and L. Legters. 1991. Distinguishing high and low anopheline-producing rice fields using remote sensing and GIS technologies. *Preventive Veterinary Medicine* 11:277–288.

Wood, B.L., L.R. Beck, R.K. Washino, S.M. Palchick, and P.D. Sebesta. 1991. Spectral and spatial characterization of rice field mosquito habitat. *International Journal of Remote Sensing* 12(3):621–626.

Wood, B.L., L.R. Beck, R.K. Washino, K.A. Hibbard, and J.S. Salute. 1992. Estimating high mosquito-producing rice fields using spectral and spatial data. *International Journal of Remote Sensing* 13(15):2813–2826.

Chapter Nine

Conclusion

This volume documents the tremendous amount of research focusing on the spatial aspect of health and disease. The copious references from the chapters attest to this dynamic and rigorous interface among spatial statistics, geographic information systems, and remote sensing. Further, more references have been brought together in the "Master GIS/RS Bibliographic Resource Guide" to provide one of the most comprehensive bibliographies on the subject. Each of the substantive chapters provides an organizing structure that functions as the "glue" that helps put together this large volume of research or "pieces" of the "puzzle."

The future of applications and research that integrate geomedical with spatial analysis, geographic information systems, and remote sensing is bright. There will be an increasing interest within this subject realm as it continues to diffuse throughout various public and private sector activities. (Geographers label this the "late majority" stage of the diffusion process.) Regardless of the growing use of these geographic technologies, the end result must translate into improving levels of human health. Our purpose, restated from the "Introduction," is to present a resource guide that might facilitate and stimulate appropriate use of geographic techniques and geographic software to health-related issues. Our hope now is that this collection, conceptualization, and synthesis of geomedical applications of spatial analysis, geographic information systems, and remote sensing will inspire others to conduct applications and research that have as their ultimate goal the betterment of health in human populations.

Master GIS/RS Bibliographic Resource Guide

D. Albert, B. Levergood, and C.M. Croner

This list of GIS/RS citations is the result of electronic databases searches, manual searches for appropriate references in existing literature, and contributions from academics and practitioners. There were numerous electronic databases available from the Health Science Library at The University of North Carolina at Chapel Hill. The key words "GIS," "Geographic(al) Information System(s)," "Remote Sensing," "Satellite Data," "Disease," and "Health" were used to limit our searches. For our purposes, the most useful electronic databases were Current Contents, MedLine 1966–, HealthStar 1975–, ERIC 1966–, and FirstSearch's OCLC (GEOBASE, GenSci Abstracts, PapersFirst, and Proceedings). Searches were made using other electronic databases (Aidsline 1980–, CINAHL 1982–, International Pharmaceutical Abstracts 1970, Core Biomedical Collection 1993–, Biomedical Collection II 1995–, Biomedical Collection III 1995–, PsycInfo 1967–, and PsycLIT 1974); however, the results were just a couple of appropriate "hits." The cutoff date for incorporating citations into our bibliographic resource guide was January 1, 1998.

The citations herein include journal articles, trade articles, and proceedings. Our guide excludes dissertations, and some articles appearing in obscure proceedings or foreign journals. Our rationale for excluding dissertations is that the best of these will find an outlet in peer-review journals. Finally, proceedings were excluded if these were difficult to find (proceedings often have a rather limited distribution). Efforts have been made to produce an up-to-date and a complete bibliographic resource; however, omissions are inevitable. For convenience, the "Master GIS/RS Bibliographic Resource Guide" is set out using subject headings. There are some 44 headings:

Aids	Leishmaniasis
Alcohol	Low Birth Weight/Infant Mortality
Asthma	Lyme Disease
Cancers	Malaria
Climate	Measles
Cholera	Methods
Contraceptives	Nursing
Dengue	Onchocerciasis
Diabetes Mellitus	Physicians

Diarrhea	Pollution
Dracunculiasis	Radon
Emergency Planning	Reviews
Encephalitis	Rift Valley Fever
Environmental Health	RMSF
Fluorosis	Ross River Virus
Foot and Mouth Disease	Sandfly Fever
Health Services Research	Schistosomiasis
Homeless	Software
Hospitals	Transportation
Immunization	Trypanosomiasis
Influenza	Tuberculosis
Lead Poisoning	Utilization

AIDS

Fost, D. 1990. Using maps to tackle AIDS. *American Demographics* 12(4):22.

Gould, P. 1997. Spreading HIV across America with an Air Passenger Operator. In *Proceedings of the International Symposium on Computer Mapping in Epidemiology and Environmental Health*, R.T. Aangeenbrug, P.E. Leaverton, T.J. Mason, and G.A. Tobin (Eds.), pp. 159–62. Alexandria, VA: World Computer Graphics Foundation.

Lam, N.S.-N. and K. Liu. 1996. Use of space-filling curves in generating a national rural sampling frame for HIV/AIDS research. *Professional Geographer* 48(3):321–322.

ALCOHOL

Millar, A.B. and P.J. Gruenewald. 1997. Use of spatial models for community program evaluation of changes in alcohol outlet distribution. *Addiction* 92(Suppl 2):S273–S283.

ASTHMA

Dunn, C.E. and S.P. Kingham. 1996. Modelling air quality and the effects on health in a GIS framework. In *Innovations in GIS: Selected papers from the Third National Conference on GIS Research UK (GISRUK)*, D. Parker (Ed.), pp. 205–214. London: Taylor & Francis.

Dunn, C.E., J. Woodhouse, R.S. Bhopal, and S.D. Acquilla. 1995. Asthma and factory emissions in northern England: Addressing public concern by combining geographical and epidemiological methods. *Journal of Epidemiology and Community Health* 49(4):395–400.

CANCERS

Bentham, G., J. Hinton, R. Haynes, A. Lovett, and C. Bestwick. 1995. Factors affecting non-response to cervical cytology screening in Norfolk, England. *Social Science and Medicine* 40(1):131–135.

Felber, G. and P.E. Leaverton. 1997. Geographic variations of childhood cancers and congenital malformations in the state of Florida. In *Proceedings of the International Symposium on Computer Mapping in Epidemiology and Environmental Health*, R.T. Aangeenbrug, P.E. Leaverton, T.J. Mason, and G.A. Tobin (Eds.), pp. 351–356. Alexandria, VA: World Computer Graphics Foundation.

Gatrell, A.C. and C.E. Dunn. 1995. Geographic information systems and spatial epidemiology: Modeling possible associations between cancer of the larynx and incineration in northwest England. In *The Added Value of Geographical Information Systems in Public and Environmental Health*, M.J.C. de Lepper, H.J. Scholten, and R.M. Stern (Eds.), pp. 215–231. Boston: Kluwer Academic.

Hjalmars, U., M. Kulldorff, G. Gustafsson, and N. Nagarwalla. 1996. Childhood leukaemia in Sweden: Using GIS and a spatial scan statistic for cluster detection. *Statistics in Medicine* 15(7/9):707–715.

Krautheim, K.R. and T.E. Aldrich. 1997. Geographic information system (GIS) studies of cancer around NPL sites. *Toxicology and Industrial Health* 13(2–3):357–362.

Lewis-Michl, E.L., J.M. Melius, L.R. Kallenbach, T.O. Talbot, M.F. Orr, and P.F. Lauridsen. 1996. Breast cancer risk and residence near industry or traffic in Nassau and Suffolk Counties, Long Island, New York. *Archives of Environmental Health* 51:255–265.

Openshaw, S., M. Charlton, C. Wymer, and A. Craft. 1987. A Mark I geographical analysis machine for the automated analysis of point data sets. *International Journal of Geographical Information Systems* 1(4):335–358.

Openshaw, S. 1996. Methods for investigating localized clustering of disease. Using a geographical analysis machine to detect the presence of spatial clustering and the location of clusters in synthetic data. *IARC Scientific Publications* 135:68–86.

Openshaw, S., A.W. Craft, M. Charlton, and J.M. Birch. 1988. Investigation of leukaemia clusters by use of a Geographical Analysis Machine. *Lancet* February 6:272–273.

Tobias, R.A., R. Roy, C.J. Alo, and H.L. Howe. 1996. Tracking human health statistics in "Radium City." *Geo Info Systems* 6(7):50–53.

Zhang, Z.-Z. and Z.-F. Zhang. 1997. The application of GIS in cancer epidemiology. In *Proceedings of the International Symposium on Computer Mapping in Epidemiology and Environmental Health*, R.T. Aangeenbrug, P.E. Leaverton, T.J. Mason, and G.A. Tobin (Eds.), pp. 211–214. Alexandria, VA: World Computer Graphics Foundation.

CLIMATE

Andrick, B., B. Clark, K. Nygaard, and A. Logar. 1997. Infectious disease and climate change: Detecting contributing factors and predicting future outbreaks. In *IGARSS '97: 1997 International Geoscience and Remote Sensing Symposium*, vol. 4, pp. 1947–1949. New York: Institute of Electrical and Electronics Engineers.

Epstein, R. 1995. Health applications of remote sensing and climate modeling. *The Earth Observer* 7(5):7–10.

CHOLERA

Collins, A. 1993. Environment and cholera in Quelimane, Mozambique: A spatial study. Occasional Paper. King's College, University of London, Department of Geography.

CONTRACEPTIVES

Entwisle, B., R.R. Rindfuss, S.J. Walsh, T.P. Evans, and S.R. Curran. 1997. Geographic information systems, spatial network analysis, and contraceptive choice. *Demography* 34(2):171–187.

Hall, G.B. and R.L. Bowerman. 1996. Using GIS to evaluate the accessibility of family planning services in the central valley of Costa Rica. *ITC Journal* 1996(1):38–48.

DENGUE

Su, M.D. and N.T. Chang. 1994. Framework for application of geographic information system to the monitoring of dengue vectors. *Kao Hsiung I Hsueh Ko Hsueh Tsa Chih [Kaohsiung Journal of Medical Sciences]* 10(Suppl):S94–S101.

DIABETES MELLITUS

Ranta, J., J. Pitkaniemi, M. Karvonen, et al. 1996. Detection of overall space-time clustering in a non-uniformly distributed population. *Statistics in Medicine* 15(2–3):2561–2572.

DIARRHEA

Emch, M. 1997. Spatial patterns of diarrheal disease in Mutlab, Bangladesh. In *Proceedings of the International Symposium on Computer Mapping in Epidemiology and Environmental Health*, R.T. Aangeenbrug, P.E. Leaverton, T.J. Mason, and G.A. Tobin (Eds.), pp. 148–153. Alexandria, VA: World Computer Graphics Foundation.

DRACUNCULIASIS

Clarke, K.C., J.P. Osleeb, J.M. Sherry, J.P. Meert, and R.W. Larsson. 1991. The use of remote sensing and geographic information systems in UNICEF's dracunculiasis (Guinea worm) eradication effort. *Preventive Veterinary Medicine* 11:229–235.

Tempalski, B.J. 1994. The case of Guinea worm: GIS as a tool for the analysis of disease control policy. *Geo Info Systems* 4(11):32–38.

World Health. 1996. Guinea worm eradication programme. *World Health* 49(3):24.

World Health. 1996. Technology aids eradication campaign. *World Health* 49(3):28.

EMERGENCY PLANNING

Coleman, D. 1994. GIS Canada: Road network partnerships paying off. *GIS World* 7(2):30.

Dunn, C.E. and D. Newton. 1992. Optimal routes in GIS and emergency planning applications. *Area* 24(3):259–267.

Feeny, M., S.K. Condon, and K. Dillman. 1997. Emergency planning through the use of GIS System. In *Proceedings of the International Symposium on Computer Mapping in Epidemiology and Environmental Health*, R.T. Aangeenbrug, P.E. Leaverton, T.J.

Mason, and G.A. Tobin (Eds.), pp. 155–158. Alexandria, VA: World Computer Graphics Foundation.

Furbee, P.M. 1995. GIS in Raleigh County: Small towns with a big database. *Journal of Emergency Medical Services* 20(6):77, 79, 81.

Grupe, F.H. 1992. Along for the ride: GIS for vehicle tracking and route selection. *Geo Info Systems* 2(8):44–47.

Marsh, S.E., C.F. Hutchinson, E.E. Pfirman, S.A. Des Rosiers, and C. van der Harten. 1994. Development of a computer workstation for famine early warning and food security. *Disasters* 18(2):117–129.

Peters, J. 1997. A GIS application: Spatial and temporal patterns of ambulance performance. In *Integrating Spatial Information Technologies for Tomorrow, GIS '97 Conference Proceedings*, pp. 49–51. Fort Collins, CO: GIS World.

Tyler, S. 1990. Computer assistance for the California earthquake rescue effort. *The Police Chief* 57(3):42–43.

Van Creveld, I. 1991. Geographic information systems for ambulance services. In *Geographic Information 1991: The Yearbook of the Association for Geographic Information*, pp. 128–130. London: Taylor & Francis.

ENCEPHALITIS

Daniel, M. and J. Kolar. 1990. Using satellite data to forecast the occurrence of the common tick *Ixodes ricinus* (L.). *Journal of Hygiene, Epidemiology, Microbiology and Immunology* 34(3):243–252.

Kitron, U., J. Michael, J. Swanson, and L. Haramis. 1997. Spatial analysis of the distribution of LaCrosse encephalitis in Illinois, using a geographic information system and local and global spatial statistics. *American Journal of Tropical Medicine and Hygiene* 57(5):469–475.

ENVIRONMENTAL HEALTH

Bowen, W.M., M.J. Salling, K.E. Haynes, and E.J. Cyran. 1995. Toward environmental justice: Spatial equity in Ohio and Cleveland. *Annals of the Association of American Geographers* 85(4):641–663.

Burke, L.M. 1993. Race and environmental equity: A geographic analysis in Los Angeles. *Geo Info Systems* 3(9):44, 46–47, 50.

Chakraborty, J. and M.P. Armstrong. 1997. Exploring the use of buffer analysis for the identification of impacted areas in environmental equity assessment. *Cartography and Geographic Information Systems* 24(3):145–157.

Chakraborty, J. and M.P. Armstrong. 1995. Using geographic plume analysis to assess community vulnerability to hazardous accidents. *Computers, Environment, and Urban Systems* 19(5–6):341–356.

Chakraborty, J. and M.P. Armstrong. 1994. Estimating the population characteristics of areas affected by hazardous materials accidents. In *GIS/LIS '94*, pp. 154–163. Bethesda, MD: American Society for Photogrammetry and Remote Sensing.

Chang, N.-B. and Y.T. Lin. 1997. Optimal siting of transfer station locations in a metropolitan solid waste management system. *Journal of Environmental Science and Health* A32(8):2379–2401.

Cutter, S.L., D. Holm, and L. Clark. 1996. Role of geographic scale in monitoring environmental justice. *Risk Analysis* 16(4):517–526.

Dearwent, S.M. and B. Hughes. 1997. The utilization of geographical information systems in describing populations proximate to hazardous substance releases. In *Proceedings of the International Symposium on Computer Mapping in Epidemiology and Environmental Health,* R.T. Aangeenbrug, P.E. Leaverton, T.J. Mason, and G.A. Tobin (Eds.), pp. 246–257. Alexandria, VA: World Computer Graphics Foundation.

De Savigny, S. and P. Wijeyaratne. 1995. *GIS for Health and the Environment: Proceedings of an International Workshop held in Colombo, Sri Lanka, 5–10 September 1994.* Ottawa, Canada: IDRC.

Dunn, C. and S. Kingham. 1996. Establishing links between air quality and health: Searching for the impossible? *Social Science and Medicine* 42(6):831–841.

Geschwind, S.A., J.A. Stolwijk, M. Bracken, E. Fitzgerald, A. Stark. C. Olsen, and J. Melius. 1992. Risk of congenital malformations associated with proximity to hazardous waste sites. *American Journal of Epidemiology* 135(11):1197–1207.

Gorynski, P., B. Wojtyniak, H. Roszkowska, I. Szutowicz, and J. Szaniecki. 1994. Studies of ambient air pollution and selected aspects of health status of children in Poznan: Preliminary information. (In Polish.) *Przeglad Epidemiologiczny* 48(3):301–305.

Hall, H.I., C.V. Lee, D. Cooper, P.A. Price-Green, and W.E. Kaye. 1997. Geographic distribution of hazardous substance releases in Iowa in 1993. In *Proceedings of the International Symposium on Computer Mapping in Epidemiology and Environmental Health,* R.T. Aangeenbrug, P.E. Leaverton, T.J. Mason, and G.A. Tobin (Eds.), pp. 258–262. Alexandria, VA: World Computer Graphics Foundation.

Lie, G.B. 1997. Tracking and investigating chemically contaminated sites using GIS. In *Proceedings of the International Symposium on Computer Mapping in Epidemiology and Environmental Health,* R.T. Aangeenbrug, P.E. Leaverton, T.J. Mason, and G.A. Tobin (Eds.), pp. 323–331. Alexandria, VA: World Computer Graphics Foundation.

Maslia, M.L., M.M. Aral, R.C. Williams, A.S. Susten, and J.L. Heitgerd. 1994. Exposure assessment of population using environmental modeling, demographic analysis, and GIS. *Water Resources Bulletin* 30(6):1025–1041.

Maroni, M., A. Fait, G. Azimonti, M. Bersani, M.G. Colombo, A. Ferioli, F. La Ferla, and F. Mainini. Programmes and activities of the International Centre for Pesticide Safety. *Central European Journal of Public Health* 3(2):103–106.

Matthies, M., F. Koormann, G. Boeije, and T.C. Feijtel. 1997. The identification of thresholds of acceptability and danger: The chemical presence route. *Archives of Toxicology* (Supplement) 19:123–35.

McMaster, R.B., H. Leitner, and E. Sheppard. 1997. GIS-based environmental equity and risk assessment: Methodological problems and prospects. *Cartography and Geographic Information Systems* 24(3):172–189.

Moragues, A. and Alcaide T. 1996. The use of a geographical information system to assess the effect of traffic pollution. *Science of the Total Environment* 189–190:267–273.

Nyerges, T., M. Robkin, and T.J. Moore. 1997. Geographic information systems for risk evaluation: Perspectives on application to environmental health. *Cartography and Geographic Information Systems* 24(3):123–144.

Richards, K. and V. Simons. 1997. Assessment of environmental and health databases for a GIS system along the U.S.–Mexico border. In *Proceedings of the International*

Symposium on Computer Mapping in Epidemiology and Environmental Health, R.T. Aangeenbrug, P.E. Leaverton, T.J. Mason, and G.A. Tobin (Eds.), pp. 332–333. Alexandria, VA: World Computer Graphics Foundation.

Runyon, T., R. Hammitt, and R. Lindquist. 1994. Buried danger: Integrating GIS and GPS to identify radiologically contaminated sites. *Geo Info Systems* 4(8):28–36.

Scott, M.S. and S.L. Cutter. 1997. Using relative risk indicators to disclose toxic hazard information to communities. *Cartography and Geographic Information Systems* 24(3):158–171.

Simons, V. and I. VanDerslice. 1997. The use of geographic information systems for incorporating environmental equity concerns into the risk assessment process. In *Proceedings of the International Symposium on Computer Mapping in Epidemiology and Environmental Health,* R.T. Aangeenbrug, P.E. Leaverton, T.J. Mason, and G.A. Tobin (Eds.), pp. 182–187. Alexandria, VA: World Computer Graphics Foundation.

Snow, S., S. Pawel, and R. Bell. 1997. SupERGIS assists in Superfund environmental restoration. *Geo Info Systems* 7(4):34–39.

Stallones, L., J.K. Berry, and J.R. Nuckols. 1992. Surveillance around hazardous waste sites: Geographic information systems and reproductive outcomes. *Environmental Research* 59(1):81–92.

Stein, A., I. Staritsky, J. Bouma, and J.W. Van Groenigen. 1995. Interactive GIS for environmental risk assessment. *International Journal of Geographical Information Systems* 9(5):509–525.

Stockwell, J.R., J.W. Sorensen, J.W. Eckert, and E.M. Carreras. 1993. The U.S. EPA geographic information system for mapping environmental releases of Toxic Chemical Release Inventory (TRI) chemicals. *Risk Analysis* 13(2):155–164.

Sui, D.Z. and J.R. Giardino. 1995. Applications of GIS in environmental equity analysis: A multi-scale and multi-zonal scheme for the city of Houston, Texas, USA. In *GIS/LIS '95,* pp. 950–959. Bethesda, MD: American Society for Photogrammetry and Remote Sensing.

Thewessen, T., R. Van de Velde, and H. Verlouw. 1992. European groundwater threats analyzed with GIS. *GIS Europe* 1(3):28–33.

Thornton, P. 1997. GIS relational database design considerations for environmental applications. In *Proceedings of the International Symposium on Computer Mapping in Epidemiology and Environmental Health,* R.T. Aangeenbrug, P.E. Leaverton, T.J. Mason, and G.A. Tobin (Eds.), pp. 334–337. Alexandria, VA: World Computer Graphics Foundation.

Valjus, J., M. Hongisto, P. Verkasalo, P. Jarvinen, K. Heikkila, and M. Koskenvuo. 1995. Residential exposure to magnetic fields generated by 110–400 kV power lines in Finland. *Bioelectromagnetics* 16(6):365–376.

Van der Veen, A.A. and A. van Beurden. 1997. Applying methods for point and area process assessment in environmental epidemiology. In *Proceedings of the International Symposium on Computer Mapping in Epidemiology and Environmental Health,* R.T. Aangeenbrug, P.E. Leaverton, T.J. Mason, and G.A. Tobin (Eds.), pp. 338–342. Alexandria, VA: World Computer Graphics Foundation.

Wagendorp, J. 1997. GIS synergisms: Linking property parcels and environmental health data. In *Proceedings of the International Symposium on Computer Mapping in Epidemiology and Environmental Health,* R.T. Aangeenbrug, P.E. Leaverton, T.J. Mason, and G.A. Tobin (Eds.), pp. 343–349. Alexandria, VA: World Computer Graphics Foundation.

Wartenberg, D., M. Greenberg, and R. Lathrop. 1993. Identification and characterization of population living near high-voltage transmission lines: A pilot study. *Environmental Health Perspectives* 101(7):626–632.

Zhang, M., S. Geng, S.L. Ustin, and K.K. Tanji. 1997. Pesticide occurrence in groundwater in Tulare County, California. *Environmental Monitoring and Assessment* 45(2):101–127.

FLUOROSIS

Chen, S. and J. Hu. 1991. Geo-ecological zones and endemic diseases in China: A sample study by remote sensing. *Preventive Veterinary Medicine* 11:335–344.

Grimaldo, M., F. Turrubiartes, J. Millan, A. Pozos, C. Alfaro, and F. Diazbarriga. 1997. Endemic fluorosis in San Luis Potosi, Mexico. III. Screening for fluoride exposure with a geographic information system. *Fluoride* 30(1):33–40.

FOOT AND MOUTH DISEASE

Sanson, R.L., H. Liberona, and R.S. Morris. 1991. The use of a geographical information system in the management of a foot-and-mouth disease epidemic. *Preventive Veterinary Medicine* 11:309–313.

HEALTH SERVICES RESEARCH

Aguglino, R. and M. Rodriguez. 1994. Evaluating public services and health assistance: Delimitation and application in a GIS. *ITC Journal* 1994(3):205–210.

Askew, P. 1993. Geographic information systems in practice: Health care planning and analysis. In *Systems d'Information Geographique et Systemes Experts*, D. Puman (Ed.), pp. 51–59. Montpellier: GIP RECLUS.

Betts, M. 1992. Make way for a better map: Travelers' desktop mapping system helps match doctors and clients. *Desktop Mapping News* 2(4):7.

Bogaerts, T. 1991. GIS for health and environment. *Cities* 8(1):17–24.

Brazil, K. and M. Anderson. 1996. Assessing health service needs: Tools for health planning. *Healthcare Management Forum* 9:22–27.

Bullen, N., G. Moon, and K. Jones. 1996. Defining localities for health planning: A GIS approach. *Social Science and Medicine* 42(6):801–816.

Clark, R. E. 1994. Cross-industry GIS problem-solving in banking and healthcare. In *GIS in Business '94*, pp. 111–120. Fort Collins, CO: GIS World.

Cliff, R. 1994. Spatial analysis in public health administration: A demonstration from WIC. In *GIS/LIS '94*, pp. 164–173. Bethesda, MD: American Society for Photogrammetry and Remote Sensing.

Curtis, S.E. and A.R. Taket. 1989. The development of geographical information systems for locality planning in health care. *Area* 21(4):391–399.

Davenhall, W. 1993. Health care reform puzzle: 1000 not-so-easy pieces. *GIS World* November/December:36–38.

Davenhall, W.F. 1993. Service geographics: The retailing of health care. In *GIS in Business '93*, pp. 195–196. Fort Collins, CO: GIS World.

Davenhall, W.F. 1995. Building healthcare geodemographic knowledge. *Business Geographics* May:46.

Davenhall, W.F. 1995. Healthcare reform: It's in the data stupid! *Business Geographics* February:48.

Davenhall, W.F. 1995. Is there a doctor in the area? *Business Geographics* July/August:34–35.

Davenhall, W.F. 1996. An all-points bulletin: Looking for healthcare data entrepreneurs. *Business Geographics* June:22.

Davenhall, W.F. 1996. Marketing Healthcare geographic technology. *Business Geographics* September:41.

Davenhall, W.F. 1996. The information therapist: Healthcare's newest profession. *Business Geographics* February:26.

Davenhall, W.F. 1997. Beating a path to better sites. *Business Geographics* February:18.

Davenhall, W.F. 1997. GIS/GPS in our personal healthcare future? *Business Geographics* July:16.

Davenhall, W.F. 1997. Market your consulting practice. *Business Geographics* May:42.

Davenhall, W.F. 1997. Spatially enabled healthcare data. *Business Geographics* September:44.

Davenhall, W.F. 1997. The data browser: A five-star management tool. *Business Geographics* April:40.

Dawson, I. and P. Aspinall. 1994. A suitable case for treatment: Using GIS and census data in the NHS. *Mapping Awareness* 8(8):24–27.

Foley, R. and P. Frost. 1996. Who cares for the carers? ...and how much does it cost? *Mapping Awareness* 10(5):28–31.

Glass, D. 1996. A World Health Organisation pilot study involving environment, public health and GIS. *Mapping Awareness & GIS in Europe* 6(9):36–40.

Gordon, A. and J. Womersley. 1997. The use of mapping in public health and planning health services. *Journal of Public Health in Medicine* 19(2):139–147.

Hale, D. 1991. The healthcare industry and geographic information systems. *Mapping Awareness* 5:36–39.

Hirschfield, A., D.J. Wolfson, and S. Swerman. 1994. Location of community pharmacies: Rational approach using geographic information systems. *International Journal of Pharmacy Practice* 3(Oct):42–52.

Howard, K. 1996. Staying on the map without losing any ground. *Earth Observation Magazine* 5(11):28–30.

Keithley, C., F. Renton, F.R. Echavarria, and R. Cantarero. 1994. Mapping the spatial patterns of health risks among minority groups in Lincoln, Nebraska. In *GIS/LIS '94*, pp. 476–484. Bethesda, MD: American Society for Photogrammetry and Remote Sensing.

Mason, K. 1994. The application of GIS to the mapping of medical data for a local health authority. *Bulletin: Society of University Cartographers* 28(1):27–35.

Nelson, J.A. 1996. Kentucky: Status of technology deployment. *Library Hi Tech* 14(2–3):131–139.

Secondini, P., L. Ciancarella, and A. Muzzarelli. 1996. GIS and public choice: The health systems case. In *Geographical Information: From Research to Application through Cooperation: Second Joint European Conference and Exhibition on Geographical Information*, M. Rumor, R. McMillan, and H.F.L. Ottens (Eds.), pp. 641–650. Amsterdam: IOS Press.

Siebert, D. 1994. The geography of healthcare reform. *GIS World* March/April:36.

Stratton, S.D. 1993. How to convince senior management of the strategic value of GIS. In *GIS in Business '93*, pp. 199–202. Fort Collins, CO: GIS World.

Terner, M. 1994. Use of GIS to support health-care marketing. In *GIS in Business '94,* pp. 121–125. Fort Collins, CO: GIS World.

Tozsa, I. and J. Galambos. 1992. Public health information system for the Erzsebetvaros district of Budapest. *Geographia Medica* 22:75–91.

Traska, M.R. 1992. Geography lessons for today's HMOs. *HMO Magazine* 33(3):36–39.

Vachon, M. 1993. GIS offers limitless potential to human service organizations. *GIS World* 6(2):52–57.

Van Teeffelen, P.B.M. and T. De Jong. 1995. Planning for services is more than drawing circles: A critical approach to rural centre planning techniques in Bantul, Indonesia. *Malaysian Journal of Tropical Geography* 26(2):151–157.

Wagendorp, J. 1995. Linked parcel/health data enhance environmental analysis. *GIS World* 8(4):54–57.

Wain, R. 1993. The use of a geographical information system in locality profiling. *Mapping Awareness and GIS in Europe* 7(8):20–22.

Warnke, J. and L.G. Jackson. 1997. GIS, health services and community development: The Quebec City English. In *Integrating Spatial Information Technologies for Tomorrow, GIS '97 Conference Proceedings,* pp. 58–60. Fort Collins, CO: GIS World.

Wrigley, N. 1991. Market-based systems of health-care provision, the NHS Bill, and geographical information systems. *Environment and Planning A* 23(1):5–8.

HOMELESS

Lee, C.-M. and D.P. Culhane. 1995. Locating the homeless: A Philadelphia case study. *Geo Info Systems* 5(7):31–34.

Lee, H. 1996. Health care for San Francisco's homeless. *Geo Info Systems* 6(6):46–47.

HOSPITALS

Furbee, P.M. and J. Spencer. 1993. Using GIS to determine travel times to hospitals. *Geo Info Systems* September:30–31.

Kohli, S., K. Sahlen, A. Sivertun, O. Lofman, E. Trell, and O. Wigertz. (1995). Distance from the primary health center: A GIS method to study geographical access to health care. *Journal of Medical Systems* 19(6):425–436.

Love, D. and P. Lindquist. 1995. The geographical accessibility of hospitals to the aged: A geographic information systems analysis within Illinois. *Health Services Research* 29(6):629–651.

Marks, A.P., G.I. Thrall, and M. Arno. 1992. Siting hospitals to provide cost-effective health care. *Geo Info Systems* 2(8):58–66.

Miller, P. 1994. Medical Center uses desktop mapping to cut costs and improve efficiency. *Geo Info Systems* 4(4):40–41.

Walsh, S.J, P.H. Page, and W.M. Gesler. 1997. Normative models and healthcare planning: Network-based simulations within a geographic information system environment. *Health Services Research* 32(2):243–260.

Walsh, S.J., W.M. Gesler, P.H. Page, and T.W. Crawford. 1995. Health care accessibility: Comparison of network analysis and indices of hospital service areas. In *GIS/LIS '95,* vol. 2, pp. 994–1005. Bethesda, MD: American Society for Photogrammetry and Remote Sensing.

Zwarenstein, M., D. Krige, and B. Wolff. 1991. The use of a geographical information system for hospital catchment area research in Natal/KwaZulu. *South African Medical Journal* 80(10):497–500.

IMMUNIZATION

Popovich, M.L. and B. Tatham. 1997. Use of immunization data and automated mapping techniques to target public health outreach programs. *American Journal of Preventive Medicine* 13(2 Suppl):102–107.

INFLUENZA

Carrat, F. and A.-J. Valleron. 1992. Epidemiologic mapping using "kriging" method: application to an influenza-like illness epidemic in France. *American Journal of Epidemiology* 135(11):1293–1300.

LEAD POISONING

Brinkmann, R., J. Lewis, J. Hagge, S. Happel, and M. Hafen. 1997. Comprehensive mapping of environmental lead data in Tampa, Florida. In *Proceedings of the International Symposium on Computer Mapping in Epidemiology and Environmental Health,* R.T. Aangeenbrug, P.E. Leaverton, T.J. Mason, and G.A. Tobin (Eds.), pp. 237–245. Alexandria, VA: World Computer Graphics Foundation.

Guthe, W.G., R.K. Tucker, E.A. Murphy, R. England, E. Stevenson, and J.C. Luckhardt. 1992. Reassessment of lead exposure in New Jersey using GIS technology. *Environmental Research* 59(2):318–325.

Hanchette, C.L. 1997. A predictive model of lead poisoning risk in North Carolina: Validation and evaluation. In *Proceedings of the International Symposium on Computer Mapping in Epidemiology and Environmental Health,* R.T. Aangeenbrug, P.E. Leaverton, T.J. Mason, and G.A. Tobin (Eds.), pp. 263–274. Alexandria, VA: World Computer Graphics Foundation.

Huxhold, W.E. 1991. *An Introduction to Urban Geographic Information Systems.* New York: Oxford University Press.

Mielke, H.W., D. Dugas, P.W. Mielke, K.S. Smith, S.L. Smith, and C.R. Gonzales. 1997. Associations between soil lead and childhood blood lead in urban New Orleans and rural Lafourche Parish of Louisiana. *Environmental Health Perspectives* 105:950–954.

Padgett, D.A. 1997. Geographic information systems techniques for delineating hot spots of childhood lead-soil exposure sources. *Proceedings of the International Symposium on Computer Mapping in Epidemiology and Environmental Health,* R.T. Aangeenbrug, P.E. Leaverton, T.J. Mason, and G.A. Tobin (Eds.), pp. 292–299. Alexandria, VA: World Computer Graphics Foundation.

Wartenberg, D. 1992. Screening for lead exposure using a geographic information system. *Environmental Research* 59(2):310–317.

Werner, R. and C. Hedlund. 1995. Putting children first: St. Paul uses GIS to prioritize lead pipe replacement. *Geo Info Systems* 5(10):44–47.

LEISHMANIASIS

Cross, E.R., W.W. Newcomb, and C.J. Tucker. 1996. Use of weather data and remote sensing to predict the geographic and seasonal distribution of *Phlebotomus papatasi* in southwest Asia. *American Journal of Tropical Medicine and Hygiene* 54(5):530–536.

LOW BIRTH WEIGHT/INFANT MORTALITY

Andes, N. and J.E. Davis. 1995. Linking public health data using geographic information system techniques: Alaskan community characteristics and infant mortality. *Statistics in Medicine* 14(5–7):481–490.

Rushton, G., R. Krishnamurthy, D. Krishnamurti, P. Lolonis, and H. Song. 1996. The spatial relationship between infant mortality and birth defect rates in a U.S. city. *Statistics in Medicine* 15(17/18):1907–1919.

Rushton, G. and P. Lolonis. 1996. Exploratory spatial analysis of birth defect rates in an urban population. *Statistics in Medicine* 15(7–9):717–726.

Rushton, G., D. Krishnamurti, R. Krishnamurthy, and H. Song. 1995. A geographic information analysis of urban infant mortality rates. *Geo Info Systems* 5(7):52–56.

Tempalski B. and S. McLafferty. 1997. Low birthweights in New York City: Using GIS to predict communities at risk. *Geo Info Systems* 7(6):34–37.

LYME DISEASE

Anon. 1995. Team tracks ticks with satellite imagery. *GIS World* July:16.

Daniel, M. and J. Kolar. 1990. Using satellite data to forecast the occurrence of the common tick *Ixodes ricinus* L. *Journal of Hygiene, Epidemiology, Microbiology and Immunology* 34(3):243–252.

De Mik, E.L., W. van Pelt, B.D. Docters-van Leewen, A. Van der Veen, J.F. Schellekens, and M.W. Borgdorff. 1997. The geographical distribution of tick bites and erythema migrans in general practice in The Netherlands. *International Journal of Epidemiology* 26(2):451–457.

Dister, S., L. Beck, B. Wood, and R. Falco. 1993. The use of GIS and remote sensing technologies in a landscape approach to the study of Lyme disease transmission risk. In *7th Annual Symposium on Geographic Information Systems in Forestry, Environment and Natural Resources Management*, pp. 1149–1156. Vancouver: Ministry of Supply and Services Canada.

Glass, G.E., B.S. Schwartz, J.M. Morgan, D.T. Johnson, P.M. Noy, and E. Israel. 1995. Environmental risk factors for Lyme disease identified with geographic information systems. *American Journal of Public Health* 85(7):944–948.

Glass, G.E., F.P. Amerasinghe, J.M. Morgan, and T.W. Scott. 1994. Predicting *Ixodes scapularis* abundance on white-tailed deer using geographic information systems. *American Journal of Tropical Medicine and Hygiene* 51(5):538–544.

Glass, G.E., J.M. Morgan, D.T. Johnson, P.M. Noy, E. Israel, and B.S. Schwartz. 1992. Infectious disease epidemiology and GIS: A case study of Lyme disease. *Geo Info Systems* 2(10):65–69.

Kitron, U. and J.J. Kazmierczak. 1997. Spatial analysis of the distribution of Lyme disease in Wisconsin. *American Journal of Epidemiology* 145(6):558–566.

Kitron, U., J.K. Bouseman, and C.J. Jones. 1991. Use of the ARC/INFO GIS to study the distribution of Lyme disease ticks in an Illinois county. *Preventive Veterinary Medicine* 11:243–248.

Kitron, U., C.J. Jones, J.K. Bouseman, J.A. Nelson, and D.L. Baumgartner. 1992. Spatial analysis of the distribution of *Ixodes dammini* (Acari: Ixodidae) on white-tailed deer in Ogle County, Illinois. *Journal of Medical Entomology* 29:259–266.

Mather, T.N., M.C. Nicholson, R. Hu, and N.J. Miller. 1996. Entomological correlates of *Babesia microti* prevalence in an area where *Ixodes scapularis* (Acari: Ixodidae) is endemic. *Journal of Medical Entomology* 33(5):866–870.

Nicholson, M.C. and T.N. Mather. 1996. Methods for evaluating Lyme disease risks using geographic information systems and geospatial analysis. *Journal of Medical Entomology* 33(5):711–720.

Serpi, T. and T.W. Forseman. 1998. Lyme disease in Maryland: Using remote sensing, GIS, and epidemiology for vector analysis and modelling. In *Proceedings for the 27th International Symposium on Remote Sensing of Environment Information for Sustainability*, pp. 805–808. Oslo, Norway: ISRSE.

MALARIA

Beck, L.R., M.H. Rodriguez, S.W. Dister, A.D. Rodriguez, E. Rejmankova, A. Ulloa, R.A. Meza, D.R. Roberts, J.F. Paris, M.A. Spanner, R.K. Washino, C. Hacker, and L.J. Legters. 1994. Remote sensing as a landscape epidemiologic tool to identify villages at high risk for malaria transmission. *American Journal of Tropical Medicine and Hygiene* 51(3):271–280.

Beck, L.R., M.H. Rodriguez, S.W. Dister, A.D. Rodrigues, R.K. Washino, D.R. Roberts, and M.A. Spanner. 1997. Assessment of a remote sensing-based model for predicting malaria transmission risk in villages of Chiapas, Mexico. *American Journal of Tropical Disease and Hygiene* 56(1):99–106.

Corbley, K.P. 1999. Identifying villages at risk of malaria spread. *Geo Info Systems* 9(1):34–38.

Corbley, K.P. 1996. Epidemiologists track virus-bearing mosquitoes. *GIS World* 9(3):52–57.

Ezzell, C. 1987. NASA mosquito watch threatened. *Nature* 325(6101):187.

Hightower, A.W. and W.A. Hawley. 1997. To map a disease: Plotting malaria's decline in western Kenya. *GPS World* 8(3):22–31.

Hightower, A.W., M. Ombok, R. Otieno, R. Odhiambo, A.J. Oloo, A.A. Lal, B.L. Nahlen, and W.A. Hawley. 1998. A geographic information system applied to a malaria field study in western Kenya. *American Journal of Tropical Medicine and Hygiene* 58(3):266–272.

Kitron, U., H. Pener, C. Costin, L. Orshan, Z. Greenberg, and U. Shalom. 1994. Geographic information system in malaria surveillance: Mosquito breeding and imported cases in Israel, 1992. *American Journal of Tropical Medicine and Hygiene* 50:550–556.

Pope, K.O., E. Rejmankova, H.M. Savage, et al. 1994. Remote sensing of tropical wetlands for malaria control in Chiapas, Mexico. *Ecological Applications* 4(1):81–90.

Rejmankova, E., D.R. Roberts, A. Pawley, S. Manguin, and J. Polanco. 1995. Predictions of adult *Anopheles albimanus* densities in villages based on distances to remotely sensed larval habitats. *American Journal of Tropical Medicine and Hygiene* 53(5):482–488.

Ribeiro, J.M., F. Seulu, T. Abose, G. Kidane, and A. Teklehaimanot. 1996. Temporal and spatial distribution of anopheline mosquitos in an Ethiopian village: Implications for malaria control strategies. *Bulletin of the World Health Organization* 74(3):299–305.

Roberts, D.R., J.F. Paris, S. Manguin, R.E. Harbach, R. Woodruff, E. Rejmankova, J. Polanco, B. Wullschleger, and L.J. Legters. 1996. Predictions of malaria vector distribution in Belize based on multispectral satellite data. *American Journal of Tropical Medicine and Hygiene* 54(3):304–308.

Roberts, D.R. and M.H. Rodriguez. 1994. The environment, remote sensing, and malaria control. *Annals of the New York Academy of Sciences* 740:396–402.

Roberts, D., M. Rodriguez, E. Rejmankova, K. Pope, H. Savage, A. Rodriguez-Ramirez, B. Wood, J. Salute, and L. Legters. 1991. Overview of field studies for the application of remote sensing to the study of malaria transmission in Tapachula, Mexico. *Preventive Veterinary Medicine* 11:269–275.

Sharma, V.P. and A. Srivastava. 1997. Role of geographic information system in malaria control. *Indian Journal of Medical Research* 106:198–204.

Sharma, V.P., R.C. Dhiman, M.A. Ansari, B.N. Nagpal, A. Srivastava, P. Manavalan, S. Adiga, K. Radhakrishnan, and M.G. Chandrasekhar. 1996. Study on the feasibility of delineating mosquitogenic conditions in and around Delhi using Indian Remote Sensing Satellite data. *Indian Journal of Malariology* 33(3):107–125.

Sharma, V.P., B.N. Nagpal, A. Srivastava, S. Adiga, and P. Manavalan. 1996. Remote sensing techniques in detection of larval habitats and estimation of larvae production. *Southeast Asian Journal of Tropical Medicine and Public Health* 27:4.

Sharp, B.L. and D. Le Sueur. 1996. Malaria in South Africa: Past, present and perspectives. (In French.) *Medecine Tropicale* 56(2):189–196.

Snow, R.W., J.R. Schellenberg, N. Peshu, D. Forster, C.R. Newton, P.A. Winstanley, I. Mwangi, C. Waruiru, P.A. Warn, C. Newbold, and K. Marsh. 1993. Periodicity and space-time clustering of severe childhood malaria on the coast of Kenya. *Transactions of the Royal Society of Tropical Medicine and Hygiene* 87(4):386–390.

Thompson, R., K. Begtrup, N. Cuamba, M. Dgedge, C. Mendis, A. Gamage-Mendis, S.M. Enosse, J. Barreto, R.E. Sinden, and B. Hogh. 1997. The Matola Malaria Project: A temporal and spatial study of malaria transmission and disease in a suburban area of Maputo, Mozambique. *American Journal of Tropical Medicine and Hygiene* 57(5):550–559.

Thomson, M.C., S.J. Connor, P. Milligan, and S.P. Flasse. 1997. Mapping malaria risk in Africa: What can satellite data contribute? *Parasitology Today* 13(8):313–318.

Thomson, M.C., S.J. Connor, P.J. Milligan, and S.P. Flasse. 1996. The ecology of malaria—As seen from Earth-observation satellites. *Annals of Tropical Medicine and Parasitology* 90(3):243–264.

Welch, J.B., J.K. Olson, M.M. Yates, A.R. Benton, and R.D. Baker. 1989. Conceptual model for the use of aerial color infrared photography by mosquito control districts as a survey technique for *Psorophora columbiae* oviposition habitats in Texas ricelands. *Journal of the American Mosquito Control Association* 5(3):369–373.

Wood, B.L., L.R. Beck, R.K. Washino, K.A. Hibbard, and J.S. Salute. 1992. Estimating high mosquito-producing rice fields using spectral and spatial data. *International Journal of Remote Sensing* 13(15):2813–2826.

Wood, B.L., L.R. Beck, R.K. Washino, S.M. Palchick, and P.D. Sebesta. 1991. Spectral and spatial characterization of rice field mosquito habitat. *International Journal of Remote Sensing* 12(3):621–626.

Wood, B., R. Washino, L. Beck, K. Hibbard, M. Pitcairn, D. Roberts, E. Rejmankova, J. Paris, C. Hacker, J. Salute, P. Sebesta, and L. Legters. 1991. Distinguishing high and low anopheline-producing rice fields using remote sensing and GIS technologies. *Preventive Veterinary Medicine* 11:277–288.

MEASLES

Solarsh, G.C. and D.F. Dammann. 1992. A community paediatric information system: A tool for measles surveillance in a fragmented health ward. *South African Medical Journal* 82(2):114–118.

METHODS

Benach, J., M.D. Garcia, and J. Donado-Caampos. 1997. GIS for mapping mortality inequalities in Spain and its socio-economic determinants constructing regions using small areas. In *Proceedings of the International Symposium on Computer Mapping in Epidemiology and Environmental Health*, R.T. Aangeenbrug, P.E. Leaverton, T.J. Mason, and G.A. Tobin (Eds.), pp. 314–322. Alexandria, VA: World Computer Graphics Foundation.

Carpenter, T.E. 1997. GIS and statistical analyses to detect temporal and spatial clustering. In *Proceedings of the International Symposium on Computer Mapping in Epidemiology and Environmental Health*, R.T. Aangeenbrug, P.E. Leaverton, T.J. Mason, and G.A. Tobin (Eds.), pp. 21–23. Alexandria, VA: World Computer Graphics Foundation.

Cox, L.H. 1996. Protecting confidentiality in small population health and environmental statistics. *Statistics in Medicine* 15(17–18):1895–905.

Cracknell, A.P. 1991. Rapid remote recognition of habitat changes. *Preventive Veterinary Medicine* 11:315–323.

Cromley, E.K., M. Cartter, and S.-H. Ertel. 1997. A method for mapping case data and population distribution to aid case-control study design. In *Proceedings of the International Symposium on Computer Mapping in Epidemiology and Environmental Health*, R.T. Aangeenbrug, P.E. Leaverton, T.J. Mason, and G.A. Tobin (Eds.), pp. 142–147. Alexandria, VA: World Computer Graphics Foundation.

Croner, C.M., L.W. White, A.A. White, and D.R. Wolf. 1992. A GIS approach to hypothesis generation in epidemiology. Technical Papers, *Proceedings: GIS and Cartography*, ASPRS/ACSM/RT92, 3:275–283.

Eddy, W.F. and A. Mockus. 1994. An example of the estimation and display of a smoothly varying function of time and space: The incidence of the disease mumps. *The Journal of the American Society for Information Science* 45(9):686–693.

Feder, J., C.M. Croner, and L.W. Pickle. 1997. Geologic and hydrologic data in epidemiologic analysis. *1996 Proceedings of the Epidemiology Section*, pp. 19–22. Chicago, IL: American Statistical Association.

Gatrell, A.C., T.C. Bailey, P.J. Diggle, and B.S. Rowlingson. 1996. Spatial point pattern analysis and its application in geographical epidemiology. *Transactions of the Institute of British Geographers* 21(1):256–274.

Gluck, M. and J. McRae. 1997. Augmented seriation: Searching for health care patterns with a multimedia cartographic tool. In *Proceedings of the International Symposium on Computer Mapping in Epidemiology and Environmental Health*, R.T. Aangeenbrug, P.E. Leaverton, T.J. Mason, and G.A. Tobin (Eds.), pp. 53–59. Alexandria, VA: World Computer Graphics Foundation.

Hayes, R.O., E.L. Maxwell, C.J. Mitchell, and T.L. Woodzick. 1985. Detection, identification, and classification of mosquito larval habitats using remote sensing scanners in earth-orbiting satellites. *Bulletin of the World Health Organization* 63(2):361–374.

Hugh-Jones, M., N. Barre, G. Nelson, K. Wehnes, J. Warner, J. Gavin, and G. Garris. 1992. Landsat-TM identification of *Amblyomma variegatum* (Acari: Ixodidae) habitats in Guadeloupe. *Remote Sensing of Environment* 40(1):43–55.

Kingham, S.P., A.C. Gatrell, and B. Rowlingson. 1995. Testing for clustering of health events within a geographical information system framework. *Environment and Planning A* 27(3):809–821.

Kirby, R.S. 1996. Toward congruence between theory and practice in small area analysis and local public health data. *Statistics in Medicine* 15(17–18):1859–1866.

Marr, P. and Q. Morris. 1997. People, poisons and pathways: A case study of ecologic fallacy. In *Proceedings of the International Symposium on Computer Mapping in Epidemiology and Environmental Health*, R.T. Aangeenbrug, P.E. Leaverton, T.J. Mason, and G.A. Tobin (Eds.), pp. 275–282. Alexandria, VA: World Computer Graphics Foundation.

Matthews, S.A. 1990. Epidemiology using a GIS: The need for caution. *Computers, Environment and Urban Systems* 14(3):213–221.

Nuckols, J.R., R.M. Reich, and C. Beseler. 1997. Use of spatial analysis in environmental health assessment. In *Proceedings of the International Symposium on Computer Mapping in Epidemiology and Environmental Health*, R.T. Aangeenbrug, P.E. Leaverton, T.J. Mason, and G.A. Tobin (Eds.), pp. 283–291. Alexandria, VA: World Computer Graphics Foundation.

Pazner, M.I. 1997. The tile map: A table/map hybrid representation in image form. In *Proceedings of the International Symposium on Computer Mapping in Epidemiology and Environmental Health*, R.T. Aangeenbrug, P.E. Leaverton, T.J. Mason, and G.A. Tobin (Eds.), pp. 105–114. Alexandria, VA: World Computer Graphics Foundation.

Pickle, L.W. and A.A. White. 1995. Effects of the choice of age-adjustment method on maps of death rates. *Statistics in Medicine* 14(5–7):615–627.

Unwin, D.J. 1996. GIS, spatial analysis and spatial statistics. *Progress in Human Geography* 20(4):540–551.

Wartenberg, D., M. Greenberg, R. Lathrop, R. Manning, and S. Brown. 1997. Using a geographic information system to identify populations living near high voltage electric power transmission lines in New York State. In *Proceedings of the International Symposium on Computer Mapping in Epidemiology and Environmental Health*, R.T. Aangeenbrug, P.E. Leaverton, T.J. Mason, and G.A. Tobin (Eds.), pp. 300–311. Alexandria, VA: World Computer Graphics Foundation.

Wittie, P.S., W. Drane, and T.E. Aldrich. 1996. Classification methods for denominators in small areas. *Statistics in Medicine* 15(17–18):1921–1926.

Wittie, P.S., T.E. Aldrich, and K. Krautheim. 1997. An integrated approach for analyzing health problems in small areas. In *Proceedings of the International Symposium on Computer Mapping in Epidemiology and Environmental Health*, R.T. Aangeenbrug, P.E.

Leaverton, T.J. Mason, and G.A. Tobin (Eds.), pp. 194–199. Alexandria, VA: World Computer Graphics Foundation.

NURSING

Begur, S.V., D.M. Miller, and J.R. Weaver. 1997. An integrated spatial DSS for scheduling and routing home-health-care nurses. *Interfaces* 27(4):35–48.

Johnston, C.L., W.R. Bischoff, G. Jeffress, and P. Michaud. 1997. The Nursing Work Force Beyond 2000 Project: The Greater Coastal Bend Region of Texas. *Journal of Nursing Administration* 27(6):4–6.

ONCHOCERCIASIS

Richards, F.O. 1993. Use of geographic information systems in control programs for onchocerciasis in Guatemala. *Bulletin of the Pan American Health Organization* 27(1):52–55.

Servat, E., J. Lapetite, J. Bader, and J. Boyer. 1990. Satellite data transmission and hydrological forecasting in the fight against onchocerciasis in West Africa. *Journal of Hydrology* 117(1–4):187–198.

PHYSICIANS

Albert, D. 1997. Monitoring physician locations with GIS. In *The National Center for Geographic Information and Analysis: GIS Core Curriculum for Technical Programs*, M. Goodchild, et al. (Eds.). Santa Barbara: University of California. (hhtp://www.ncgia.ucsb.edu/education/curricula/cctp/applications/albert.html).

Albert, D. 1995. Is there a doctor near the house? MapInfo analyzes health care access in North Carolina. *GlobalNews*, Troy, NY: MapInfo Corporation. (Summer):7.

Hirschfield, A., P. Brown, and P. Bundred. 1993. Doctors, patients and GIS: A spatial analysis of primary health care provision on the Wirral. *Mapping Awareness & GIS in Europe* 7(9):9–12.

Jacoby, I. 1991. Geographic distribution of physician manpower: The GMENAC legacy. *Journal of Rural Health* 7(4 Suppl):427–426.

Jankowski, P. and G. Ewart. 1996. Spatial decision support for health practitioners: Selecting a location for rural health practice. *Geographical Systems* 3:279–299.

Prabhu, S. 1995. A generalized framework for information systems for physician recruitment/referral. *Journal of Management in Medicine* 9(4):24–30.

POLLUTION

Blaha, D., B. Mosher, R.C. Harriss, C.E. Kolb, B. McManus, J.H. Shorter, and B. Lamb. 1994. Mapping urban sources of atmospheric methane. *Geo Info Systems* 4(10):34–38.

Briggs, D.J., S. Collins, P. Elliot, P. Fischer, S. Kingham, E. Lebret, K. Pryl, H. Van Reeuwijk, K. Smallbone, and A. Van der Veen. 1997. Mapping urban air pollution using GIS: A regression-based approach. *International Journal of Geographical Information Science* 11(7):669–718.

Cade, S.H. and R.D. Stephens. 1994. Remote sensing of vehicle exhaust emissions. *Environmental Science and Technology* 28:258A.

Cambridge, H., S. Cinderby, J. Kuylenstierna, and M.J. Chadwick. 1996. "A hard rain's gonna fall..." Environmental modelling of acid rain. *GIS Europe* 5(2):20–22.

Chwastek, J. and T.Z. Dworak. 1990. Satellite remote sensing of industrial air pollution in the Cracow special protected area. *Journal of Environmental Pathology, Toxicology and Oncology* 10(6):288–289.

Corwin, D.L., P.J. Vaughan, and K. Loague. 1997. Modeling nonpoint source pollutants in the vadose zone with GIS. *Environmental Science and Technology* 31(8):2157–2175.

Garnsworthy, J. 1992. Department of the Environment: GIS activities. *Mapping Awareness & GIS in Europe* 6(4):42–45.

Gates, N.S. and R. Spellicy. 1992. Remote sensing offers new look at air pollution. *Control and Systems* 39(3):16.

Henriques, W.D. and K.R. Dixon. 1996. Estimating spatial distribution of exposure by integrating radiotelemetry, computer simulation, and geographic information system (GIS) techniques. *Human and Ecological Risk Assessment* 2:527–538.

Jacquez, G.M. 1995. The map comparison problem: Tests for the overlap of geographic boundaries. *Statistics in Medicine* 14:2343–2361.

Liu, L.-J.S., R. Delfino, and P. Koutrakis. 1997. Ozone exposure assessment in a southern California community. *Environmental Health Perspectives* 105(1):58–65.

Moore, T.J. 1993. Assessing the risk from air toxics. *Geo Info Systems* 3(2):42–50.

Pikhart, H., V. Prikazsky, M. Bobak, B. Kriz, M. Celko, J. Danova, K. Pyrl, and J. Pretel. 1997. Association between ambient air concentrations of nitrogen dioxide and respiratory symptoms in children in Prague, Czech Republic. Preliminary results from the Czech part of the SAVIAH study. Small Area Variation in Air Pollution and Health. *Central European Journal of Public Health* 5(2):82–85.

Sengupta, S. and P. Venkatachalam. 1994. Health hazard assessment in an industrial town with the help of GRAM-GIS. *Environmental Monitoring and Assessment* 32(2):155–160.

RADON

Coleman, K.A., G.C. Hughes, and E.J. Scherieble. 1994. Where's the radon? The geographic information system in Washington State. *Radiation Protection Dosimetry* 56(1–4):211–213.

Fitzpatrick-Lins, K., T.L. Johnson, and J.K. Otton. 1990. Radon potential defined by exploratory data analysis and geographic information systems. *U.S. Geological Survey Bulletin* 1908:E1–E10.

Geiger, C. and K.B. Barnes. 1994. Indoor radon hazard: A geographical assessment and case study. *Applied Geography* 14(4):350–371.

Kohli, S., K. Sahlen, O. Lofman, A. Sivertun, M. Foldevi, E. Trell, and O. Wigertz. 1997. Individuals living in areas with high background radon: A GIS method to identify populations at risk. *Computer Methods and Programs in Biomedicine* 53(2):105–112.

Siniscalchi, A.J., S.J. Tibbetts, R.C. Beakes, X. Soto, M.A. Thomas, N.W. McHone, and S. Rydell. 1996. A health risk assessment model for home owners with multiple pathway radon exposure. *Environmental International* 22(Suppl. 1):S739–S747.

REVIEWS

Albert, D. 1994. Geographic information systems (GIS). In *Geographic Methods for Health Services Research: A Focus on the Rural-Urban Continuum*, T.C. Ricketts, L.A. Savitz, W.M. Gesler, and D.N. Osborne (Eds.), pp. 201–206. Lanham, MD: University Press of America.

Albert, D. 1997. Synopsis and bibliographic resource of medical-GIS applications. In *The National Center for Geographic Information and Analysis: GIS Core Curriculum for Technical Programs*, M. Goodchild et al. (Eds.). Santa Barbara: University of California.(hhttp://www.ncgia.ucsb.edu/education/curricula/cctp/applications/med_bibliography.html).

Albert, D.P, W.M. Gesler, and P.S. Wittie. 1995. Geographic information systems and health: An educational resource. *Journal of Geography* 94(2):350–356.

Albert, D. and B. Levergood. 1997. Pointers to the literature on health applications of remote sensing. *World Health Forum* 18(1):88–91.

Andrick, B. 1997. GIS & AI in the analysis & prediction of disease outbreaks. *Proceedings of Small College Computing Annual Symposium*, pp. 335–343. SCCS.

Anon. 1996. Use of geographic information systems in epidemiology (GIS-Epi). *Epidemiological Bulletin* 7(1):1–6.

Arambulo, P.V. and V. Astudillo. 1991. Perspectives on the application of remote sensing and geographic information system to disease control and health management. *Preventive Veterinary Medicine* 11:345–352.

Barandela, R. 1997. Geographic information systems and environmental assessment: Difficulties and opportunities. *ITC Journal* 1997(1):74–78.

Barinaga, M. 1993. Satellite data rocket disease control efforts into orbit. *Science.* 261(5117):31–32.

Barnes, C.M. 1991. An historical perspective on the applications of remote sensing to public health. *Preventive Veterinary Medicine* 11:163–166.

Barnes, S. and A. Peck. 1994. Mapping the future of health care: GIS applications in health care analysis. *Geo Info Systems* 4(4):30–39.

Beck, L.R., B.L. Wood, and S.W. Dister. 1995. Remote sensing and GIS: New tools for mapping human health. *Geo Info Systems* 5(9):32–37.

Briggs, D.J. and P. Elliot. 1995. The use of geographical information systems in studies on environment and health. *World Health Statistics Quarterly* 48(2):85–94.

Chen, C., U.S. Tim, and S. Stratton. 1996. Application of GIS in Environmental Epidemiology: Assessment of Progress and Future Trends. In *GIS/LIS '96*, pp. 853–869. Bethesda, MD: American Society for Photogrammetry and Remote Sensing.

Clarke, K.C., S.L. McLafferty, and B.J. Tempalski. 1996. On epidemiology and geographic information systems: A review and discussion of future directions. *Emerging Infectious Diseases* 2(2):85–92.

Cline, B.L. 1970. New eyes for epidemiologists: Aerial photography and other remote sensing techniques. *American Journal of Epidemiology* 92(2):85–89.

Colwell, R.R. 1996. Global climate and infectious disease: The cholera paradigm. *Science* 274(5295):2025–2031.

Cowen, D.J. 1995. The importance of GIS to the average person. In *GIS in Government: The Federal Perspective, 1994*, pp. 7–11. Fort Collins, CO: GIS World.

Croner, C.M., J.S. Sperling, and F.R. Broome. 1996. Geographic information systems (GIS): New perspectives in understanding human health and environmental relationships. *Statistics in Medicine* 15(17–18):1961–1977.

De Lepper, M.J.C., H.J. Scholten, and R.M. Stern. 1995. *The Added Value of Geographical Information Systems in Public and Environmental Health.* Boston: Kluwer Academic Publishers.

Epstein, P.R., D.J. Rogers, and R. Slooff. 1993. Satellite imaging and vector-borne disease. *The Lancet* 341(8857):1404–1406.

Gatrell, A.C. and T.C. Bailey. 1997. Can GIS be made to sing and dance to an epidemiological tune? In *Proceedings of the International Symposium on Computer Mapping in Epidemiology and Environmental Health,* R.T. Aangeenbrug, P.E. Leaverton, T.J. Mason, and G.A. Tobin (Eds.), pp. 38–52. Alexandria, VA: World Computer Graphics Foundation.

Glass, G.E., J.L. Aron, J.H. Ellis, and S.S. Yoon. 1993. *Applications of GIS Technology to Disease Control.* Baltimore: The Johns Hopkins University, School of Hygiene and Public Health, Department of Population Dynamics.

Glushko, E.V., S.M. Malkhazova, and V.S. Tikunov. 1991. Space monitoring of spread and dynamics of natural-focus diseases (as evidenced by plague in Kyzyl-Kum). *Geographia Medica* 21:77–88.

Gobalet, J.G. and R.K. Thomas. 1996. Demographic data and geographic information systems for decision making: The case of public health. *Population Research and Policy Review* 15(5–6):537–548.

Hay, S.I. 1997. Remote sensing and disease control: Past, present and future. *Transactions of the Royal Society of Tropical Medicine and Hygiene* 91(2):105–106.

Hay, S.I., M.J. Packer, and D.J. Rogers. 1997. The impact of remote sensing on the study and control of invertebrate intermediate hosts and vectors for disease. *International Journal of Remote Sensing* 18(14):2899–2930.

Hightower, A.W. and R.E. Klein. 1995. Building a geographical information system (GIS) public health infrastructure for research and control of tropical diseases. *Emerging Infectious Diseases* 1(4):156–157.

Hugh-Jones, M. 1989. Applications of remote sensing to the identification of the habitats of parasites and disease vectors. *Parasitology Today* 5(8):244–251.

Hugh-Jones, M. 1991. Introductory remarks on the application of remote sensing and geographic information systems to epidemiology and disease control. *Preventive Veterinary Medicine* 11:159–161.

Jovanovic, P. 1987. Remote sensing of environmental factors affecting health. *Advances in Space Research* 7(3):11–18.

Jovanovic, P. 1988. Satellite technology for primary health care. *World Health Forum* 9(3):386–387.

Jovanovic, P. 1989. Satellite medicine. *World Health* January–February:18–9.

Jovanovic, R. 1991. Adopting remote sensing for public health. *Preventive Veterinary Medicine* 11:357–358.

Kellar, A.A. and V.I. Kuvakin. 1997. New problems of military geography. (In Russian.) *Voenno-Meditsinskil Zhurnal* 318(4):14–8, 80.

Kingman, S. 1989. Remote sensing maps out where the mosquitoes are. *New Scientist* 123(1682):38.

Kitron, U. 1997. Surveillance of vector-borne diseases: Role of GIS, Remote Sensing and Spatial Analysis. In *Proceedings of the International Symposium on Computer Mapping in Epidemiology and Environmental Health,* R.T. Aangeenbrug, P.E. Leaverton, T.J. Mason, and G.A. Tobin (Eds.), pp. 163–74. Alexandria, VA: World Computer Graphics Foundation.

Malkhazova, S.M. and P.V. Petrov. 1994. The monitoring of health condition of population Russia. In *Intercarto: GIS for Environmental Studies and Mapping Conference*, pp. 111–112. Fort Collins, CO: GIS World.

Malkazova, S.M. and V.S. Tikunov. 1994. Analysis of spatial differences in the health of the population of the world (with the example of natural-endemic diseases). In *Intercarto: GIS for Environmental Studies and Mapping Conference*, pp. 105–110. Fort Collins, CO: GIS World.

Mott, K.E., I. Nuttall, P. Desjeux, and P. Cattand. 1995. New geographical approaches to control of some parasitic zoonoses. *Bulletin of the World Health Organization* 73(2):247–257.

Nuttall, I. 1997. GIS—Management tools for the control of tropical diseases: Applications in Botswana, Senegal and Morocco. In *Proceedings of the International Symposium on Computer Mapping in Epidemiology and Environmental Health*, R.T. Aangeenbrug, P.E. Leaverton, T.J. Mason, and G.A. Tobin (Eds.), pp. 175–181. Alexandria, VA: World Computer Graphics Foundation.

Openshaw, S. 1996. Geographical information systems and tropical diseases. *Transactions of the Royal Society of Tropical Medicine and Hygiene* 90(4):337–339.

Pan American Health Organization. 1996. Use of geographic information systems in epidemiology (GIS-Epi). *Epidemiological Bulletin* 17(1):1–6.

Rochon, G.L. 1995. Food security, public health, and sustainable development: Participation of U.S. agencies and the United Nations in remote sensing and GIS in Africa. In *Integrating Spatial Information Technologies for Tomorrow, GIS '97 Conference Proceedings*, pp. 52–57. Fort Collins, CO: GIS World.

Scholten, H.J. and M.J. de Lepper. 1991. The benefits of the application of geographical information systems in public and environmental health. *World Health Statistics Quarterly* 44(3):160–170.

Stephenson, J. 1997. Ecological monitoring helps researchers study disease in environmental context. *JAMA* 278(3):189–191.

Taylor, D. 1997. Seeing the forests for more than the trees. *Environmental Health Perspectives* 105(11):1186–1191.

Tim, U.S. 1997. GIS in environmental and public health sciences: Opportunities, practical issues, and future trends. In *Integrating Spatial Information Technologies for Tomorrow, GIS '97 Conference Proceedings*, pp. 52–57. Fort Collins, CO: GIS World.

Tim, U.S. 1995. The application of GIS in environmental health sciences: Opportunities and limitations. *Environmental Research* 71(2):75–88.

Travis, J. 1997. Spying diseases from the sky: satellite data may predict where infectious microbes will strike. *Science News* 152(5):72–73.

Twigg, L. 1990. Health based geographical information systems: Their potential examined in the light of existing data sources. *Social Science and Medicine* 30(1):143–155.

Vine, M.F., D. Degnan, and C. Hanchette. 1997. Geographic information systems: Their use in environmental epidemiologic research. *Environmental Health Perspectives* 105(6):598–605.

Waller, L.A. 1996a. Epidemiologic uses of geographic information systems. *Statistics in Epidemiology Report* 3(1):1–7.

Waller, L.A. 1996b. Geographic information systems and environmental health. *Health and Environmental Digest* 9(10):85–88.

Washino, R.K. and B.L. Wood. 1994. Application of remote sensing to arthropod vector surveillance and control. *American Journal of Tropical Medicine and Hygiene* 50(Suppl. 6):134–144.

RIFT VALLEY FEVER

Ambrosia, V.G., K.G. Linthicum, C.L. Bailey, and P. Sebesta. 1989. Modeling Rift Valley fever (RVF) disease vector habitats using active and passive remote sensing systems. In *IGARSS '89 Remote Sensing: An Economic Tool for the Nineties*, 2758–60. Vancouver: IGARSS '89 12th Canadian Symposium on Remote Sensing.

Davies, F.G., E. Kilelu, K.J. Linthicum, and R.G. Pegram. 1992. Patterns of Rift Valley fever activity in Zambia. *Epidemiology and Infection* 108(1):185–191.

Pope, K.O., E.J. Sheffner, K.J. Linthicum, C.L. Bailey, T.M. Logan, E.S. Kasischke, K. Birney, A.R. Njogu, and C.R. Roberts. 1992. Identification of central Kenyan Rift Valley fever virus vector habitats with Landsat TM and evaluation of their flooding status with airborne imaging radar. *Remote Sensing of Environment* 40(3):185–196.

Linthicum, K.J., C.L. Bailey, F.G. Davies, and C.J. Tucker. 1987. Detection of Rift Valley fever viral activity in Kenya by satellite remote sensing imagery. *Science* 235(4796):1656–1659.

Linthicum, K.J., C.L. Bailey, C.J. Tucker, K.D. Mitchell, T.M. Logan, F.G. Davies, C.W. Kamau, P.C. Thande, and J.N. Wagateh. 1990. Application of polar-orbiting, meteorological satellite data to detect flooding of Rift Valley fever virus vector mosquito habitats in Kenya. *Medical and Veterinary Entomology* 4(4):433–438.

Linthicum, K.J., C.L. Bailey, C.J. Tucker, D.R. Angleberger, T. Cannon, T.M. Logan, P.H. Gibbs, and J. Nickeson. 1991. Towards real-time prediction of Rift Valley fever epidemics in Africa. *Preventive Veterinary Medicine* 11:325–334.

Silberner, J. Rift Valley fever: Long-distance diagnosis. 1987. *Science News* 131(28 March):199.

ROCKY MOUNTAIN SPOTTED FEVER (RMSF)

Cooper, J.W. and J.U. Houle. 1991. Modeling disease vector habitats using thematic mapper data: Identifying *Dermacentor variabilis* habitat in Orange County, North Carolina. *Preventive Veterinary Medicine* 11:353–354.

ROSS RIVER VIRUS

Dale, P.E.R. and C.D. Morris. 1996. *Cules annulirostris* breeding sites in urban areas: Using remote sensing and digital image analysis to develop a rapid predictor of potential breeding areas. *Journal of the American Mosquito Control Association* 12(2 pt. 1):316–320.

SANDFLY FEVER

Cross, E., W.W. Newcomb, and C.J. Tucker. 1996. Use of weather data and remote sensing to predict the geographic and seasonal distribution of *Phlebotomus papatasi* in southwest Asia. *American Journal of Tropical Medicine and Hygiene* 54(5):530–536.

Cross, E., C.J. Tucker, and K.C. Hyams. 1997. The use of AVHRR and weather data to detect the seasonal and geographic occurrence of *Phebotomus papatasi* in Southwest Asia. In *Proceedings of the International Symposium on Computer Mapping in Epidemiology and Environmental Health*, R.T. Aangeenbrug, P.E. Leaverton, T.J.

Mason, and G.A. Tobin (Eds.), pp. 24–26. Alexandria, VA: World Computer Graphics Foundation.

SCHISTOSOMIASIS

Abdel-Rahman, M.S., M.M. el-Bahy, N.M. el-Bahy, and J.B. Malone. 1997. Development and validation of a satellite based geographic information system (GIS) model for epidemiology of Schistosoma risk assessment on snail level in Kafr El-Sheikh Governorate. *Journal of the Egyptian Society of Parasitology* 27(2):299–316.

Cross, E.R., C. Sheffield, R. Perrine, and G. Pazzaglia. 1984. Predicting areas endemic for schistosomiasis using weather variables and a Landsat data base. *Military Medicine* 149(10):542–544.

Malone, J.B., M.S. Abdel-Rahman, M.M. Elbahy, O.K. Huh, M. Shafik, and M. Bavia. 1997. Geographic information systems and the distribution of schistosoma mansoni in the Nile delta. *Parasitology Today* 13(3):112–119.

Malone, J.B., O.K. Huh, D.P. Fehler, P.A. Wilson, D.E. Wilensky, R.A. Holmes, and A.I. Elmagdoub. 1994. Temperature data from satellite imagery and the distribution of schistosomiasis in Egypt. *American Journal of Tropical Medicine and Hygiene* 50(6):714–722.

Zhou, Y., D. Maszzle, P. Gong, R. Spear, and X. Gu. 1996. GIS based spatial network models of Schistosomiasis infection. *Geographic Information Sciences* 2(1–2):51–57.

SOFTWARE

Aldrich, T., K. Krautheim, E. Kinee, D.J. Wanzer, and D. Tibara. 1997. Statistical methods for space-time cluster analyses. In *Proceedings of the International Symposium on Computer Mapping in Epidemiology and Environmental Health*, R.T. Aangeenbrug, P.E. Leaverton, T.J. Mason, and G.A. Tobin (Eds.), pp. 226–236. Alexandria, VA: World Computer Graphics Foundation.

Evans, J. 1997. The lay of the land: GIS and mapping software. *Health Management Technology* 18(5):58, 60.

Gatrell, A.C. and T.C. Bailey. 1996. Interactive spatial data analysis in medical geography. *Social Science and Medicine* 42(6):843–855.

Hall, H.I., C.V. Lee, and W.E. Kaye. 1996. Cluster: A software system for epidemiologic cluster analysis. *Statistics in Medicine* 15(7–9):943–950.

Nobre, F.F., A.L. Braga, R.S. Pinheiro, and J.A. Lopes. 1997. GISEpi: A simple geographical information system to support public health surveillance and epidemiological investigations. *Computer Methods and Programs in Biomedicine* 53(1):33–45.

Patterson, D. 1995. Mapping out your future (desktop computers in nursing home management). *Nursing Homes* 44(8):34–35.

Ruiz, M. First impressions: Health statistics mapping software. 1996. *Geo Info Systems* 6(6):52–55.

TRANSPORTATION

Austin, K., M. Tight, and H. Kirby. 1997. The use of geographical information systems to enhance road safety analysis. *Transportation Planning and Technology* 20(3):249–266.

Austin, K. 1995. The identification of mistakes in road accident records: Part 1, locational variables. *Accident Analysis and Prevention* 27:261–276.

Lepofsky, M., M. Abkowitz, and P. Cheng. 1993. Transportation hazard analysis in an integrated GIS environment. *Journal of Transportation Engineering* 119(2):239–254.

Braddock, M., G. Lapidus, E. Cromley, R. Cromley, G. Burke, and L. Banco. 1994. Using a geographic information system to understand child pedestrian injury. *American Journal of Public Health* 84(7):1158–1161.

Cromley, E.K. and G.M. Lapidus. 1995. Surveillance of child pedestrian injuries: A GIS approach. *Data Needs in an Era of Health Reform: Proceedings of the 25th Public Health Conference on Records and Statistics and the National Committee on Vital Health Statistics 45 Anniversary Symposium,* pp. 97–101. Washington, DC, Hyattsville, MD: U.S. Department of Health and Human Services, Public Health Service, Centers for Disease Control and Prevention, National Center for Health Statistics.

D'Arcy, W.J. 1995. Pennsylvania DOT maps local accidents. *Geo Info Systems* 5(5):34–35, 69.

Jones, A. 1993. Using GIS to link road accident outcomes with health service accessibility. *Mapping Awareness & GIS in Europe* 7(8):33–37.

Jones, A.P., I.H. Langford, and G. Bentham. 1996. The application of K-function analysis to the geographical distribution of road traffic accident outcomes in Norfolk, England. *Social Science and Medicine* 42(6):879–885.

Lovett, A.A., J.P. Parfitt, and J.S. Brainard. 1997. Using GIS in risk analysis: A case study of hazardous waste transport. *Risk Analysis* 17(5):625–633.

Padgett, D.A. 1992. Assessing the safety of transportation routes for hazardous materials. *Geo Info Systems* 2(2):46–48.

TRYPANOSOMIASIS

Brady, J. 1991. Seeing flies from space. *Nature* 351:695.

Clark, P.A. 1997. TB or not TB? Increasing door-to-door response to screening. *Public Health Nursing* 14(5):268–271.

Reid, R.S., C.J. Wilson, R.L. Kruska, and W. Mulatu. 1997. Impacts of tsetse control and land-use on vegetative structure and tree species composition in south-western Ethiopia. *Journal of Applied Ecology* 34(3):731–747.

Robinson, T., D. Rogers, and B. Williams. 1997. Univariate analysis of tsetse habitat in the common fly belt of southern Africa using climate and remotely sensed vegetation data. *Medical and Veterinary Entomology* 11(3):223–234.

Rogers, D.J. 1991. Satellite imagery, tsetse and trypanosomiasis in Africa. *Preventive Veterinary Medicine* 11:201–220.

Rogers, D.J. and B.G. Williams. 1993. Monitoring trypanosomiasis in space and time. *Parasitology* 106(Suppl):S77–S92.

Rogers, D.J. and S.E. Randolph. 1991. Mortality rates and population density of tsetse flies correlated with satellite imagery. *Nature* 351(6329):739–741.

Rogers, D.J., S.I. Hay, and M.J. Packer. 1996. Predicting the distribution of tsetse flies in West Africa using temporal Fourier processed meteorological satellite data. *Annals of Tropical Medicine and Parasitology* 90(3):225–241.

TUBERCULOSIS

Beyers, N., R.P. Gie, H.L. Zietsman, M. Kunneke, J. Hauman, M. Tatley, and P.R. Donald. 1996. The use of a geographical information system (GIS) to evaluate the distribution of tuberculosis in a high-incidence community. *South African Medical Journal* 86(1):40–44.

UTILIZATION

Feinleib, M. 1997. The use of computer mapping in monitoring the Nation's Health. In *Proceedings of the International Symposium on Computer Mapping in Epidemiology and Environmental Health*, R.T. Aangeenbrug, P.E. Leaverton, T.J. Mason, and G.A. Tobin (Eds.), pp. 1–3. Alexandria, VA: World Computer Graphics Foundation.

Gould, M.I. 1992. The use of GIS and CAC by health authorities: Results from a postal questionnaire. *Area* 24(4):391–401.

Heywood, I. 1990. Geographic information systems in social sciences. *Environment and Planning A* 22(7):849–854.

Mohan, J. and D. Maguire. 1985. Harnessing a breakthrough to meet the needs of health care. *Health and Social Service Journal* May 9:580–81.

Nicol, J. 1991. Geographic information systems within the National Health Service: The scope of implementation. *Planning Outlook* 34(1):37–42.

Glossary

Compiled by C.M. Croner, W.M. Gesler, and D.P. Albert

absolute space: Space considered as a container which may or may not be filled with objects; associated with Euclidean geometry where points are related by fixed distances or metrics between them.

activity spaces: The local areas within which people move or travel in the course of their daily activities.

Advanced Very High Resolution Radiometer (AVHRR): A five-channel scanning device that quantitatively measures electromagnetic radiation.

band: The relatively slender section of the electromagnetic spectrum that a remote sensor can discern.

boundary: A line that defines the limits of a geographic entity such as a block, BNA, census tract, county, or place.

Cartesian coordinate system: Points, lines, and polygons are most commonly defined on maps using x, y Cartesian coordinates such as latitude/longitude based on principles of Euclidean geometry.

CBD: Central Business District.

census block group: A combination of census blocks that is a subdivision of a census tract or BNA.

census block numbering area (BNA): An area delineated by State officials or (lacking State participation) by the Census Bureau, following Census Bureau guidelines, for the purpose of grouping and numbering decennial census blocks in counties or statistically equivalent entities in which census tracts have not been established.

census block: The smallest entity for which the Census Bureau collects and tabulates decennial census information.

census designated place (CDP): A statistical entity, defined for each decennial census according to Census Bureau guidelines, comprising a densely settled concentration of population that is not within an incorporated place, but is locally identified by a name.

census tract: A small, relatively permanent statistical subdivision of a county in a metropolitan area or a select non-metropolitan county, for presenting decennial census data.

central place theory: Based on ideas of city and town spacing by Walter Christaller, the theory deals with hierarchies of goods and services, threshold populations required to market these goods and services, and ranges or territories over which the goods and services are sold. As used here, the theory applies to the distribution of health manpower or facilities within an urban area. Large hospitals or specialist physicians are higher order functions and thus require larger thresholds and ranges.

centroid: The central location within a specified geographic area.

choropleth mapping: Shows data value for a predefined area such as a census tract (or watershed) or statistically defined Thiessen polygon, where the entire area is shaded to a representative summary statistic.

cluster analysis: A method of grouping spatial units or variables measured over spatial units by bringing together the two "closest" units or variables, the next closest units or variables and so on until all units or variables are in one cluster. The number of clusters finally chosen for further analysis is based on maximizing between cluster differences and minimizing within cluster differences.

cluster dendogram: The tree-like diagram output from a cluster analysis that indicates how the spatial units or variables in the analysis are related to each other in clusters.

coefficient of areal correspondence: This measure, defined as the ratio of the intersection of two phenomena to the union of the same two phenomena can be used to determine the degree of correspondence between disease maps and maps of possible factors related to diseases.

connectivity: The connectivity of a point or vertex of a graph or network is the degree to which the point is linked to other points. It can be measured in several ways.

contagious diffusion: Diffusion of an idea, innovation or disease based on person-to-person contact within a local population, depending to a great extent on the distances among people.

contiguous: Descriptive of geographic areas that are adjacent to one another, sharing either a common boundary or point.

correlation coefficient: A measure of the strength of the relationship between two interval data variables.

Delauney triangles: The process of converting point data into contours that uses a triangulated network which joins neighboring points together. Interpolation of values observed at point locations produces an approximation of the distribution.

delineate: To draw or identify on a map the specific location of a boundary.

difference maps: A method of analyzing the pattern of differences between two maps which are based on the same areal units by comparing the number of units where they are the same or different with a theoretical, random distribution.

Digital Line Graph (DLG) and Digital Elevation Model (DEM): Two major U.S. Geological Survey (USGS) digital data file structures. DLG is line map information in digital form. These files include information on planimetric base categories, such as transportation, hydrography, and boundaries. DEM files consist of a sample array of elevations for a number of ground positions that are at regular spaced intervals.

digital number: This is the brightness value for each pixel.

digital orthophotos: These are digital photographs or "photo maps" that result from processing aerial photographs to remove image distortion and displacement due to perspective, camera tilt and terrain relief.

directed graph: A graph or network in which relations among points or vertices are either unequal and reciprocal or non-reciprocal.

dispersion (of a graph): A measure of the total distance among nodes or vertices in a graph or network, the sum of the rows of the matrix of shortest paths among points.

eccentricity: A measure of the shape of an ellipse, defined as the ratio of the distance between the foci and the length of the major axis. Smaller values indicate greater ellipse elongation or "out-of-roundness."

ecological fallacy: Most often referred to as a logical flaw which results from making a causal inference about an individual phenomenon or process on the basis of observations on groups.

electromagnetic radiation: The range of electromagnetic radiation wavelengths and frequencies (i.e., radio, microwave, infrared, visible, ultraviolet, X-ray, and gamma-ray radiation).

empirical Bayes (EB) mapping: A parametric statistical procedure to stabilize statistics, prior to mapping, by Bayesian modelling which "shrinks" the statistics from areas with a small population toward an overall mean.

feature: Any part of the landscape, whether natural (such as a stream or ridge) or artificial (such as a road or power line). In a geographic context, features are any part of the landscape portrayed on a map, including legal entity boundaries such as city limits or county lines.

FIPS (Federal Information Processing Standards) code: One of a series of codes, issued by the National Institute of Standards and Technology (NIST), assigned for the purpose of ensuring uniform identification during computer processes involving geographic entities throughout all Federal Government programs and agencies.

flow analysis: The analysis of flows (e.g., patients) among various nodes or vertices (e.g., homes and hospitals) of a network.

Fourier series: Expressions used to fit trend surfaces to spatially distributed variables if the variables are thought to behave in an oscillatory manner across space.

free sampling: When calculating Moran's join count measures, the assumption that the probabilities of obtaining a "black" or "white" area are known in advance.

geocoding: Digital procedure for finding map coordinates that correspond to data attributes of features. For example, address geocoding is the ability to reference a street address or street intersection to a location on the map. The Census Bureau's TIGER system provides a national computer-readable map database for geocoding operations.

geodetic control: A network of surveyed and monumented points on the earth's surface whose locations are established in accordance with national accuracy standards. A state plane coordinate system comprises a system of x, y coordinates for each state and is commonly used in GIS.

geometric corrections: The process by which points in an image are registered to corresponding points on a map or another image that has already been rectified.

geographic base file (GBF): A generic term for a computer file of geographic attributes of an area (such as street names, address ranges, geographic codes, hydrography, railroads).

global positioning system (GPS): Set of twenty-four NAVSTAR GPS satellites orbiting 12,000 miles above the Earth that are constantly transmitting a precise time and their position in space. GPS receivers (on or near the ground) use the transmitted signals to triangulate a position.

graph theory: A theoretical framework for examining the relationships or links (represented by lines) that exist among places, towns, regions and so on (represented by points, nodes or vertices).

grid: Provides graph theory relationship principles to mapping by use of a grid mesh to define a regular but arbitrary polygon framework for "holding" geographic data. The grid technique inherently involves association with a coordinate system, but it does not necessarily require precise association.

hierarchical clustering: A method which emphasizes how adjacent spatial units with high or low disease rates might cluster by ranking the units by disease rate and then examining how probable cluster adjacencies would be compared to random conditions and marking off successive clusters wherever low probability values occur.

hierarchical diffusion: A diffusion process in which an idea, innovation or disease spreads by moving from larger to smaller places, often with little regard to the distance between places.

Internet (and related terms): A worldwide system for linking smaller computer networks together. Internet connected networks use a particular set of communication standards to communicate, known as TCP/IP. The World Wide Web (WWW) is a hypertext system which links images, sounds, and text, collectively known as hypermedia. A user navigates the WWW, travelling from page to page, and browses information via mouse on text and images. Uniform Resource Locator (URL) is an address on the World Wide Web.

isoline: A line connecting points of equal value. For example, isotherms connect points of equal temperature and isomorbs connect points with equal disease rates.

kernel estimator: Relates to a class of density estimators (that avoid dependence on essentially arbitrary spatial units) called "kernel methods" from which to derive a smoothing estimator or parameter; "adaptive" kernel estimation provides that the bandwidth parameter should be spatially variable to account for variation in density in the study region.

kriging: This mapping method represents the variable under study as a continuous process, unconstrained by the borders of geographic units and where sudden transitions between levels of two neighboring areas are avoided.

Landsat: Land Remote-Sensing Satellite.

latency: The period between the onset of any given cause and disease detection.

linear feature: A feature, such as a railroad, road, street, stream, pipeline, or boundary that can be represented by a line in a geographic data base.

linear transect: A line drawn across an area along which samples for data analysis are chosen.

location quotients: A method of spatial standardization which can be used to map disease or health care resources.

Lorenz curves: A graphical representation of spatial inequality in health care resources which plots cumulative resources like medical manpower against cumulative populations for spatial units.

Master Address File (MAF): The Census Bureau's permanent list of addresses for individual living quarters that is linked to the TIGER database and will be continuously maintained through partnerships with the U.S. Postal Service, and Federal, State, regional and local agencies, and the private sector.

mean center: The mean center of a set of points in space is a unique point whose x- and y-coordinates are the means of the x- and y-coordinates of the individual points.

metric properties: Those properties of a relation defined on a space that require a fixed distance among points in the space. Nonmetric properties do not have this requirement.

metropolitan area (MA): A collective term, established by the Federal OMB and used for the first time in 1990, to refer to metropolitan statistical areas (MSAs), consolidated metropolitan statistical areas (CMSAs), and primary metropolitan statistical areas (PMSAs).

modifiable units: The idea that if the boundaries of the spatial units one deals with are changed, the results of spatial analysis will differ also.

moisture index: The MI is calculated using Landsat TM imagery as (band 4 – band 7)/(band 4 + band 7).

Monte Carlo methods: Techniques for estimating the solution of a numerical or mathematical problem by means of an artificial sampling experiment.

multidimensional scaling: Presents the structure of distance-like data as a geometrical picture.

nearest neighbor analysis: A technique of determining whether a set of points in space is distributed in a regular, random or clustered pattern by comparing the mean distance of points to their nearest neighbors to the mean distance expected if the pattern were random.

nodality: The nodality of a point or vertex of a graph or network is the number of network paths or links incident at that point.

nonparametric spatial smoothers: Mathematical methods of smoothing data spatially that are not dependent on parametric statistical distribution, for example, nearest-neighbor estimators, Markov point processes.

normalized difference vegetation index: The NDVI is computed generally as follows: (near infrared – red)/(near infrared + red).

pathogenesis: Cellular and other events and reactions and other pathologic mechanisms occurring in the development of disease.

pixel: picture element.

Poisson distribution or process: A spatial point process in which each quadrat (grid cell area) in an area has an equal and independent chance of containing a point.

polygon: A closed, two-dimensional figure with three or more sides and intersections. For example, a polygon could be represented as an enclosed geographic area such as a land parcel or political jurisdiction.

population potential: The potential at any point on a population potential surface is the sum of the reciprocals of distance (or some power of distance) of every person in the population from the point. It is population density weighted by the reciprocal of distance and is analogous to magnetic or electric fields.

power series polynomials: Polynomial expressions of linear, quadratic, cubic or higher orders that are used in the fitting of trend surfaces to spatially distributed variables.

quadrat analysis: A method of determining whether a pattern of points is clustered, random or regular by comparing an observed frequency distribution of points in grid cells with an expected (Poisson) frequency distribution if the distribution were random.

radiometric resolution: This refers to the sensitivity of the sensor to incoming radiation.

random walk: A method of simulating random movements by beginning at a point and proceeding to a succession of new points by choosing angles at random and moving a standard distance in the directions indicated by these angles.

region (census geographic): Four groupings of States (Northeast, South, Midwest, and West) established by the Census Bureau in 1942 for the presentation of census data. Each region is subdivided into divisions.

relational database model: Database structure where each record is a "row" and each field is a "column." A set of rows stored under columns comprise a "table." For graphical objects, the graphical information and corresponding coordinates are attached to each row in the relational database table.

relative risk (ratio): The ratio of the incidence of a disease among those exposed to the disease to the incidence among those not exposed.

relative space: Space as a relation defined on a set of objects, without the requirement of fixed distances or metrics among points in the space.

remote sensing: The analysis and interpretation of the earth's landscape and resources using aerial photography or satellite imagery. This is especially useful for public health analysis in the study of disease host and vector habitats, extent, and magnitude of disease events and natural disasters, and changes in observations over time.

residuals (from trend surface models): The differences between observed values at points and values predicted by a trend model, an indication of over- or under-prediction at various points.

resolutions: Characteristics of remote sensing data that include spatial, spectral, radiometric, and temporal resolutions.

slope, aspect: The term slope addresses the steepness of an area while the aspect is related to the direction in which the area is oriented.

spatial autocorrelation: The degree of relatedness of a set of spatially located data; the extent to which adjoining or neighboring spatial units influence particular variables recorded on those units.

spatial correlogram: A graph which records the degree of spatial autocorrelation among a set of areal units at different spatial lags where the first lag involves adjoining or neighboring units, the second lag neighbors of neighbors, and so on.

spatial resolution: The measurement of the minimum distance between two objects that will allow them to be differentiated from one another in an image.

spatial stochastic process: Formalizes the way in which spatial association is generated or generally expresses how observations at each location depend on values at neighboring locations, that is, on the spatial lags. As with time series analysis, spatial stochastic processes can be classified as spatial autoregressive (SAR) or spatial moving average (SMA) processes.

spectral resolution: The number and size of the bands recorded by a sensor determine the instrument's spectral resolution.

standard deviational ellipse: An ellipse whose major and minor axes are drawn to represent the magnitude of the minimum and maximum dispersion of a set of points from their mean center.

standard distance: A measure of the dispersion of a set of points in space, analogous to the standard deviation of a set of data.

standardized mortality ratio: The number of deaths, either total or cause-specific in a given sub-population or spatial unit expressed as a percentage of the number of deaths that would have been expected in that sub-population or unit if the age- and sex-specific rates in the general population had obtained.

state economic area (SEA): A group of adjacent counties within a State that have similar economic and social characteristics, as determined by various governmental agencies. A SEA may be a single metropolitan county with unique characteristics.

suitability analysis: A variant of spatial analysis in which GIS map layers are integrated to form a composite choropleth map for decision making. Often used in site selection, it results from an application of one or more mathematical relations (functions, transformations) to the attributes of one or more maps. Examples are maps derived from weighted intersection overlay and weighted multidimensional scaling.

Summary Tape File (STF): One of a series of computer files containing large amounts of decennial census data for the various levels of the Census Bureau's geographic hierarchy.

TIGER database: A computer file that contains geographic information representing: the position of roads, rivers, railroads, and other census-required map features; attributes associated with each feature, such as feature name, address ranges, and class codes; position of the boundaries for those geographic areas that the Census Bureau uses in its data collection, processing, and tabulation operations; and attributes with those areas, such as their names and codes. This file is stored in multiple partitions, for example, counties or groups of counties, although it represents all U.S. space (including Puerto Rico, and the Outlying Areas) as a single seamless data inventory.

TIGER/Line® files: public extracts of selected geographic and cartographic information from the Census Bureau's TIGER (Topologically Integrated Geographic Encoding and Referencing) database.

topography: The collective features on the surface of the earth, including relief, hydrography and cultural features. Topographic maps of the United States are produced by the U.S. Geological Survey for GIS uses. In public health, topographic maps can reveal disease associations with elevation, surface water, wind direction, solar exposure, time, and other factors.

topology: One component of the science of mathematics dealing with geometric configurations that do not vary when transformed through bending, stretching, or mapping at various scales.

trend surface analysis: The decomposition of each observation of a spatially distributed variable into components associated with regional and local effects. Fitting a function which relates values of points in space to the point coordinates to create a trend surface, and examination of the residuals.

Triangulated Irregular Network (TIN): A contouring method of linear interpolation that bases its predicted surface on the flat plane that can be fitted to any three non-collinear points.

variance/mean ratio test: A student's t-test which compares the ratio of the variance and mean of a histogram of quadrat (grid cell) counts of a set of points in space to the expected ratio of 1 if the points were randomly distributed.

velocity field: A graphical representation of journey times between pairs of points in an urban area.

Voronoi or Theissen Polygons: Method of mathematically transforming point data into thematic maps based not on predetermined reporting units but rather on proximity of the distribution of points. One of the points which define the edges of a Theissen polygon is always the nearest neighbor to the point in the center of the polygon.

ZIP (Zone Improvement Plan) Code: A five-, seven-, nine-, or eleven-digit code assigned by the U.S. Postal Service to a section of a street, a collection of streets, an establishment, structure, or group of post office boxes, for the delivery of mail.

Subject Index

Geographical Index

A

Africa. *See also* individual countries by name
  central 26
  malaria statistics 119
  sub-Saharan 111
Akron, Ohio, measles incidence 26
Alameda County, California 47
Alberta, Canada 21
Argentina, breast cancer mortality 22
Arizona
  Phoenix 15
  Yuma County 67
Asia
  China 44
  malaria statistics 119
Australia, Wollongong 16–17

B

Benin, Zou Province, dracunculiasis 112, 116, 155
Brazil, Braganca Paulista County 14, 24
Buffalo, New York, surface water
      contamination 88

C

California
  AIDS, spread of 22
  Alameda County 47
  environmental databases 81
  Lyme disease 116
  San Francisco 47, 61, 67–68
  toxic emissions 48
  western malaria mosquito 157
Canada
  Alberta 21
  Toronto, Ontario 20
Central Africa 26
Charlotte, North Carolina 64–66
China 44

D

Denton County, Texas, landfill citing 90
Des Moines, Iowa
  birth defects 30
  infant mortality/birth defects 24, 27

Duluth, Minnesota, hospital market share
      62–63
Durham County, Great Britain 15

E

East Orange, New Jersey 48
Egypt, moisture variability and RS 158
England. *See also* Great Britain; United
      Kingdom
  fowl pest disease 14
  Lancashire 14–15
  leukemia clusters 15
  Nottingham 16, 18
  Portsmouth 48
  southwestern 17
Ethiopia 15, 124
Evans County, Georgia, variables and
      cardiovascular death 27

F

Fairfax County, Virginia, radon exposure 49
Flint, Michigan 16
Florida
  AIDS, spread of 22
  herbicide toxicity 91
France 15

G

Georgia
  Evans County 27
  Savannah 16, 28
Germany, Munich 16, 20, 21
Great Britain. *See also* England; United
      Kingdom
  childhood leukemia 15
  fowl pest disease 14
  Northumberland and Durham Counties 15
  Wales 14, 18, 84–85
Groton, Massachusetts, ground water
      contamination 87
Guadeloupe, bont tick habitat 157, 159, 161

H

Henan, China, esophageal cancer 44

215

T - #0243 - 111024 - C0 - 254/178/11 - PB - 9780367578947 - Gloss Lamination